Scorpions

Plus Other Popular Invertebrates

By Jerry G. Walls

LCCN: 96-183295
ISBN: 1-882770-86-2

Cover photo by Carol Polich (front), and Paul Freed (back).
The additional photographs in this book are by Paul Freed, pp. 5, 6, 15, 37, 38, 42, 43, 44, 52, 55, 59, 63–65, 71, 72, 73, 75, 77 (top), 78 (bottom), 79–82; James Gerholdt, pp. 8, 9, 12, 14, 17, 25, 50, 53, 61, 69, 77 (bottom), 78 (top); Bill Love, pp. 21, 58; Maleta M. Walls, pp. 27, 30, 32, 33.

2030 Main Street, Suite 1400
Irvine, CA 92614
www.facebook.com/luminamediabooks
www.luminamedia.com

CONTENTS

CHAPTER 1

INTRODUCING SCORPIONS

Scorpions are among the strangest of all animals on the land, and they are instantly recognizable by everyone. Their flattened form, oval body, slender "tail," and large "claws" give them a vague resemblance to crayfish and to other crustaceans; actually, they are arachnids, close relatives of spiders and mites. Like tarantulas and other spiders, all scorpions are venomous, which means that all have venom glands as well as the ability to inject venom into prey or predator. But unlike spiders—whose venom generally is not fatal to an adult human—there are quite a few scorpion species (certainly more than twenty) that are able to kill humans with their sting.

Why would anyone want to keep such a venomous animal as a pet? The reality is that many species of scorpions are attractive (in their own fashion), easy to keep, and actually harmless or nearly so—especially if not handled at all. They are not pets in the manner of a cat or dog, and the wise keeper will never try to stroke a scorpion or let it take a walk around the room, but they are fun to observe and certainly have their supporters. Currently, about a hundred thousand scorpions of various types from around the world are offered for sale each year in the United States, Europe, and Japan. Scorpions may be unusual pets, but they certainly are not rare pets.

Where Scorpions Fit

Scorpions have jointed legs and a segmented body and thus are placed in the phylum Arthropoda along with the

With the tail held high over the body and pincers outstretched, scorpions are one of the most recognizable animals.

crustaceans, insects, millipedes, and similar invertebrates. The presence of four pairs of walking legs (the pedipalps at the front of the body that bear the pincers are not true legs, as we will discuss shortly), the absence of antennae, and the lack of jaws place scorpions in the class Arachnida, along with the spiders, harvestmen, mites and ticks, and even stranger animals known as whip scorpions and wind scorpions. There are estimated to be more than seventy thousand named species of arachnids; almost half are spiders, and the other half are mites, which leaves little room for the rest of the arachnid groups.

Order Scorpiones

The order Scorpiones holds all the scorpions, living and fossil. Scorpions are distinguished from other arachnids by the long, segmented postabdomen ("tail") at the end of the body that ends in a curved sting, and by the presence of unique sensory organs, the pectines, under the body in both sexes. Currently, scientists who study scorpions (sometimes called scorpiologists but more likely to think of themselves as arachnologists, scientists who study arachnids in general) recognize more than fifteen hundred species of living scorpions, plus roughly forty genera of fossils. Animals recognized as scorpions have been around since the Silurian Period (four hundred million years ago) and were among the first animals to walk on land. The most primitive fossil scorpions definitely were water animals, complete with gills, though they probably looked much like modern scorpions with the usual large pincers and sting. Scorpions are considered among the most primitive of the arachnids, having a poorly formed nervous system compared with spiders. In fact, scorpions show many similarities to the even more primitive marine horseshoe crabs (*Limulus* and relatives) that are themselves relicts from ancient days.

Families of Scorpions

For decades, it was traditional to say that there were six scorpion families that held all the genera and species of living scorpions. These were Bothriuridae, Buthidae, Chactidae, Diplocentridae, Scorpionidae, and Vaejovidae. In the 1980s, arachnologists began more closely studying the many types of scorpions and decided that these six families gave a false impression of how the scorpions were related to each other. So, they began to rearrange the genera and set up new families to better represent true relationships. This led at one point to recognizing nine families by 1990, and then sixteen families by 2000. Today, there is an ongoing tendency to merge some of these families; at the time of writing, many arachnologists recognize thirteen scorpion families (see following chart).

This reorganization has led to some confusion—especially among beginning hobbyists who are new to the nomenclature. Several of the thirteen families are of techni-

This black scorpion, *Parabuthus transvaalicus*, may be called by one of several common names, including spitting thick-tailed scorpion and South African fat-tailed scorpion. A member of family Buthidae, it is one of the more toxic species and packs a seriously painful sting.

cal interest only as they contain very rare and poorly known scorpions. In addition, this reorganization and recent merging of families has led to some of the most common genera of pet scorpions being placed in unfamiliar families.

To give you a better idea of your scorpion's relatives, the following chart lists the thirteen families, familiar generic names mentioned in the hobby, and some technical notes.

Family	Notes	Familiar Genera
Bothriuridae	More than 100 species.	Bothriurus, Timogenes
Buthidae	The largest family of scorpions, with almost 700 species. All the dangerously venomous scorpions belong to this family.	*Androctonus, Babycurus, Buthacus, Buthus, Centruroides, Hemibuthus, Hottentotta* (formerly *Buthotus*), *Isometrus, Leiurus, Lychas, Parabuthus, Rhopalurus, Tityus*
Caraboctonidae	Though this "new" family has fewer than 20 species, it includes the giant hairy scorpions of the North American deserts and their relatives from South America.	*Hadrurus, Hadruroides*
Chactidae	Almost impossible to define, this family holds about 150 species.	*Anuroctonus, Brotheas, Uroctonus*
Chaerilidae	Only 1 Asian genus, *Chaerilus*, with about 20 species.	*Chaerilus*
Euscorpiidae	Though a small family with only some 65 species, the major scorpions of southern Europe are included here.	*Euscorpius, Megacormus, Scorpiops*
Hemiscorpiidae	About 70 species.	*Hadogenes, Hemiscorpius, Iomachus, Opisthacanthus*
Iuridae	Once a larger family, this now contains just 2 species from Turkey and Greece.	*Iurus*
Microcharmidae	The 6 species in this family were all described since 1995 and are less than an inch (2.5 cm) long.	*Microcharmus, Neoprotobuthus*
Pseudochactidae	The single species of this doubtfully distinct family was described from Asia in 1998.	*Pseudochactas*
Scorpionidae	More than 200 species now are placed in this family, including those that were traditionally held in the former family Diplocentridae. The largest and heaviest scorpions are in this family.	*Diplocentrus, Heterometrus, Nebo, Opistophthalmus, Pandinus, Scorpio*
Superstitioniidae	Most of the 10 species in this family are found in caves, but some are burrowers in deserts.	*Superstitionia, Typhlochactas*
Vaejovidae	Many of the roughly 150 species of this family are found in the deserts of the western United States.	*Paruroctonus, Serradigitus, Smeringurus, Vaejovis*

Scorpion Anatomy

Though scorpions range in length from less than half an inch (12.5 mm) to 7 inches (18 cm) and from slender to heavy-bodied, all scorpions are built along identical lines.

External Appearance

The covering of a scorpion's entire body—the exoskeleton—is a thick, inflexible, and nonabsorbent skin called the cuticle, which usually is hardened by calcium salts. Narrow rings of thinner cuticle (membranes) that retain their flexibility allow the different body parts to move. The entire exoskeleton of a scorpion is shed during the molt (see chapter 5), including the internal lining of the gut and book lungs and the complicated structures inside the legs that allow motion. The body consists of three obvious parts:

- a large cephalothorax (or prosoma) that houses the bases of the legs and pedipalps as well as the mouthparts and eyes;
- the wide preabdomen (or mesosoma) that consists of seven segments, the last segment distinctly wider at its base than at its rear end;
- and a slender postabdomen (or metasoma) commonly called the "tail," or cauda, that consists of five narrow, elongated rings before ending in a bulbous telson (not a true segment) that contains the venom glands and the sting (or aculeus).

Pandinus species, such as the specimen shown here, are some of the largest scorpions available today, making them quite popular in the pet trade.

With this frontal view of an emperor scorpion, we can see the mouthparts and chelicerae.

The upper surface of the cephalothorax is a single large shield called the carapace. In addition to grooves and ridges that sometimes are useful in identification, the major features of the carapace are the pair of large, usually black eyes near the center of the shield and two to five pairs of small eyes at each front corner. The eyes can detect light and dark and perhaps movement, but they are not believed to be able to distinguish shapes. The eyes perhaps help the scorpion keep track of daily and seasonal changes in day length and thus help coordinate mating. The number of small eyes varies somewhat with family, genus, and species; and in many of the scorpions that live in caves, these eyes may be entirely absent (as may the large central or major eyes).

Under the cephalothorax are the bases of the walking legs and pedipalps, the sternum where the legs originate, and the openings to the reproductive tracts. The sternum is a conspicuous plate lying at the bases of the third and fourth legs. In almost all scorpions, the sternum is a roughly five-sided (pentagonal) plate; however, in the family Buthidae it is distinctly triangular, and in the family Bothriuridae it is a narrow horizontal strip. Immediately behind the sternum is a round or oval plate, or pair of plates, called the genital operculum, which hides the opening to the female's reproductive tract or to the male's pair of penis lobes.

Just behind the genital operculum is the most unusual organ found in scorpions, the pectines. The pectines (plural, also pectines) consists of two wing-like structures (each called a pecten), each with a row of teeth on its back edge. Arachnologists recognize many other parts that make up the pectines as well, though most parts are visible and countable only under magnification. Anyone can see the teeth, however, on the pectines of even a small scorpion.

Pectines seem to serve two purposes: 1) to chemically sense the presence of other scorpions, especially of the opposite sex; and 2) in males, to help determine if the substrate is solid enough to support a sperm packet. As a rule, male scorpions have more slender pecten teeth, and more of them, than do females of the same species—though the counts and sometimes shapes can be the same or similar.

The dorsal plates (tergites) of the preabdomen generally are simple straps, but commonly they carry raised ridges or groups of small tubercles in distinct patterns. On the plates' undersides, you can see four pairs of spiracles (or stigma) that are openings to the book lungs, through which scorpions breathe. The spiracles usually can be tightly closed by special muscles to prevent the entry of water and toxic gases; a scorpion can "hold its breath" for many minutes without harm.

The postabdomen (or tail) always consists of five rings (or segments) plus the telson bearing the sting. The width of the rings varies greatly from species to species, as does the ornamentation with crests and tubercles. In many male scorpions, the fourth and fifth rings are elongated compared with those of the female. The telson may be greatly swollen or relatively slender, the sting nearly straight or strongly curved, and there often is a tubercle, or tooth (the subaculear tubercle), under the base of the sting. The anus of a scorpion is in the soft membrane between the lower back edge of the fifth ring and the base of the telson.

Appendages

Let's get the walking legs out of the way first. There are four pairs of walking legs in all scorpions, and they are amazingly similar in structure throughout the group. Starting at the base (under the body of the scorpion), the seven parts of each leg are: the coxa, which joins to the sternum; the trochanter, visible at the side of the body; the long femur; a swollen patella below the bend of the leg; a moderately long tibia; a tarsus divided into two parts, the basitarsus (tarsomere I) and teleotarsus (tarsomere II); and finally, a very short, hidden pretarsus carrying three claws, the central one very small and hard to see. Commonly, there is a large, pointed spine at the end of the tibia (the tibial spur) and one or two spines at the end of tarsomere I (the pedal, or tarsal spurs). Usually, there also are many spines and enlarged bristles (setae) under tarsomere II that help the scorpion burrow, walk through shifting sand, or hang onto bark when climbing.

The fifth, or front, pair of "legs" in a scorpion are the pedipalps, which end in pincers. The pedipalp's coxa is hidden under the body, but the trochanter usually is large and visible. It is followed by a long femur and then a swollen patella. The pincers are the hand (or manus) of the scorpion, consisting of an inflated basal segment (palm) and immovable finger formed from the tibia and the movable finger, or tarsus. Pincer shape varies greatly, from quite slender with long fingers to gigantically inflated palms with short, thick fingers studded with crushing teeth. Found on the segments of the pedipalps are special thick setae, or bristles, with movable bases; these are called trichobothria and occur in distinct patterns that vary not only with each family, but also with each species. Like other bristles on the legs of a scorpion, trichobothria are easily lost after death and may not be easy to see without magnification.

The final appendages of the scorpion are the pair of clawlike chelicerae (singular, chelicera) at the very front end of the cephalothorax. These have a hidden basal segment and large, immovable and movable fingers much as in the pedipalp pincer. There are strong teeth on the fingers that are used to help catch prey and to rip it into smaller parts for easy digestion. Unlike spiders, which have venom glands in their chelicerae, there are no such glands in scorpion chelicerae—though they have the ability to bite strongly and may even draw blood.

Sting

The telson lies at the end of the tail and is not considered a true body segment like the rings of the tail; instead, it is an outgrowth of the last ring. Within the telson are two large venom glands and muscles that can press against them to force out venom. The venom, a complex and variable mixture of proteins, exits through a pair of openings (one for each gland) just before the end of the sting. Stinging is under voluntary control in scorpions; they can release large or small amounts of venom, or none at all (dry stings).

Internal Anatomy

Like most invertebrates, a scorpion has an open circulatory system. That is, no distinct blood vessels lead from the heart to the internal organs, as occurs in the closed circulatory systems of vertebrates and a limited number of invertebrates; instead, copper-based blood (vertebrate blood is iron based) called hemolymph bathes the organs. The heart is a long cylinder running down the center of the back just under the tergites of the preabdomen, from just behind the carapace to the base of the tail. In each side of the heart are seven openings, or ostia, through which blood is pumped at low pressure into the body cavity. A single artery leads to the brain, which is an enlargement of the nerve cord around the base of the gut, below the central pair of eyes. The gut is a long cylinder with many side pouches (called ceca) that

Shown here is an emperor scorpion's sting, covered in many sensory bristles that it uses to get a constant readout of the environment around itself.

runs from between the bases of the chelicerae to the end of the fifth tail ring. Other internal organs—including the complicated, many-branched testes of the male and the ovariuterus of the female—are housed in the preabdomen.

As mentioned above, scorpions have an open circulatory system; blood runs into and bathes every part of the animal to keep the tissues alive and also functions similar to fluid in a hydraulic system by powering the legs and other body parts to move. Also bathed in blood are the book lungs—four pairs—that open on the undersurface of the preabdomen through the spiracles. Each book lung consists of layers and folds of thin tissue (hence the name "book" lung) that allow oxygen from the air to seep into the blood and carbon dioxide to release from the body. Breathing is passive, without any obvious abdominal movements.

Some Natural History

Scorpions are found on all the continents except Antarctica. They seldom are found in cool climates, though they range throughout southwestern Canada in North America and in cooler regions of Australia and South America. The majority of the species are found in tropical and subtropical habitats of Central and South America, Africa, and southern Asia; few scorpions are found on the Pacific Islands, and only a few range north into central Europe. They occupy habitats that range from humid rain forests to the driest deserts, from below sea level in western North America and near the Red Sea to high altitudes in mountains of Asia and South America.

All scorpions are nocturnal (active at night), generally hiding during the day in a burrow or under debris. Most scorpions are terrestrial, seldom climbing except in low shrubs; however, many species of the family Buthidae are very good climbers and may actively chase prey high into the trees or even across the ceilings of homes.

Scorpions lack true jaws and thus begin digestion outside the body. The scorpion uses its pedipalps to grab the prey (all scorpions are carnivores, taking only insects and other animal prey, usually living) and transfer it to the chelicerae.

Here, the prey is chopped and pulled into smaller parts that are ground between the strong bases of the chelicerae and pedipalps to release the prey animal's body fluids. The scorpion then covers the prey in digestive juices. The scorpion's mouth is a small opening under the center of the front edge of the cephalothorax that is protected by setae of various types and sizes that help strain out large bits of prey and only allow the fluids to be sucked into the gut for full digestion. When the scorpion has gotten as much fluid as possible from the food, the hard parts are dropped. Digestion is relatively rapid, since the food is liquid, and the feces are just small, dry particles.

Scorpions may be either ambush predators that wait in the mouths of their burrows for prey to come close enough to attack, or active hunters that roam about at night trying to find sleeping or inactive insects, lizards, frogs, and even small mammals. Scorpions will feed on other scorpions as well as spiders, and females may kill and eat their mates and (rarely) their own young.

Scorpions are usually solitary and may be cannibalistic, preying on smaller individuals, including their own siblings. Many exceptions to this statement exist; bark scorpions, for example, may occur in large colonies and wintering masses where the animals are jammed next to each other in a hiding place and don't seem to mind. Many emperor and forest scorpions won't attack each other, at

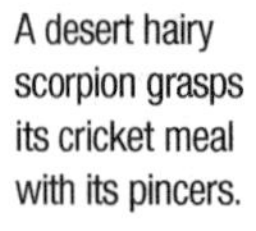

A desert hairy scorpion grasps its cricket meal with its pincers.

Mother scorpions carry their young on their backs for several weeks.

least when adult, but they may eat young of their own species. As a general rule, it is never safe to house two scorpions together for any length of time; otherwise, you are likely to end up with just one scorpion.

All scorpions give live birth; the female holds the embryos within her body until they are fairly well developed. The mother carries her young on her back for several weeks until they can live on their own. Contrary to some statements, mother scorpions seldom eat their young before the young have assumed their own lives on the substrate, at which point they become fair game. One exception to this is the giant hairy scorpion; in many cases, at least in captive conditions, a mother eats her young that are riding on her back.

As a rule, smaller scorpions seem to live relatively short lives, perhaps only three to five years, whereas the larger species may live seven to fifteen years. We actually know very little about the natural life spans of scorpions, though some keepers have been able to maintain species of *Pandinus* and *Heterometrus* (the largest emperor and forest scorpions, which have relatively slow development) in the vivarium for nearly twenty years.

Species that are quickest to mature, such as many small buthids (*Centruroides* species, *Tityus* species), may live only three years in nature yet start producing young when only a few months old. Most larger scorpions take one to three years to sexually mature.

CHAPTER 2

SELECTION

Selecting a good pet scorpion has two phases: first, finding what looks like a healthy animal; and second, deciding whether it is safe to keep a particular species. Because of the possibility of escapes and accidents, beginners are especially encouraged to keep only "safe" species.

Recognizing a Healthy Scorpion

You are unlikely to find any captive-bred scorpions offered for sale. Almost all are collected from the wild, often at night by using ultraviolet lights (black lights), which make the scorpions "glow" (see the following section, Self-Collecting). Others may be farmed in their native countries—that is, the babies are hatched from gravid wild females and exported to a purchaser in a different country. Almost all scorpions offered for sale are adults of uncertain age or are at least large young with only a few molts to go before becoming adult. There is no way to know whether a specimen is two years old or ten years old.

A healthy scorpion will be active under low-light conditions and will retreat when challenged. If a pencil is slowly waved in front of it, it will either try to escape to its burrow or shelter, or it will attack, touching the pencil to determine if it might be food. A scorpion that just lies on its belly and moves little or not at all is certainly ill.

Healthy scorpions usually hold the tail high over the body, with the sting ready for use. The pedipalps are held out, the trichobothria feeling air movements and helping the scorpion determine if it is sharing its space with other animals. Bark scorpions (*Centruroides*) hold the tail horizontally alongside the body so it does not get in the way as the scorpion climbs; this is perfectly normal.

When inspecting potential purchases, take note of whether the scorpion can hold its tail erect. Choose an active specimen that holds its tail high over the body. A listless scorpion is not likely to be in good health.

Scorpions sometimes are offered for sale with many specimens housed together in a single small container; this is very stressful to the animals. In nature, scorpions seldom are social (with the exception of *Centruroides* and their relatives). If you are interested in a particular specimen, ask the dealer to move it to a separate container and give it a few minutes to adjust. Then, offer it a cricket or mealworm and see how it reacts. If the scorpion is healthy, it should attack the food or at least notice it is present, even if not hungry. Randomly choosing a single specimen from a mass of scorpions in a cage is not the way to get a healthy animal.

No Regeneration

Putting the specimen in a separate container also allows you to observe its body. Are all the legs intact with each ending in a claw? Are both pedipalps complete, with functioning pincers? A scorpion without good pedipalps probably will starve because it cannot catch enough prey. The chelicerae should look intact, with both fingers complete, though this may be hard to see. Check the obvious as well—does the tail end in a telson with a complete sting? Sometimes, stings may be broken during collecting or shipment, and wild-caught scorpions may be found missing a telson. Though a scorpion can survive well in captivity without a sting and venom glands (crickets are not much of a challenge), you want an intact specimen.

What about regeneration? Tarantulas and lizards regrow missing parts, such as legs and tails, so why not scorpions? Forget it. Unless the scorpion is very young and still actively molting (which is unlikely in wild-caught specimens offered for sale), what you see is what you get, even if the scorpion lives another dozen years. A broken leg or tail may heal completely, producing a dark brown scar at its end, but there have been few instances of even partial regeneration of missing appendages in older scorpions. Most scorpiologists believe that adult scorpions don't even molt, so there is no chance of regrowing a missing part.

Stress

Many scorpions offered for sale may be stressed, though there are few outward signs to indicate stress. Stress is due not only to crowding (which most scorpions seem to forget about as soon as they are moved to new quarters), but also to dehydration and too cool or hot temperatures during shipment. It is thought that desert scorpions get all or almost all their water requirements from their food, but emperor and forest scorpions often drink. Fortunately, many dehydrated scorpions will be fine if allowed to drink their fill during their first few days in a new vivarium.

Scorpions become inactive in low temperatures, but most can survive temperatures below 60°F (16°C) for long periods and even near-freezing temperatures for a short while without long-term effects. Heat is much more dangerous to a scorpion. The potential for heat-induced stress is best determined by looking at the container in which the scorpions are offered for sale. Occasionally, I've seen a dozen or more big *Pandinus* scorpions offered in a 5-gallon (19-liter) tank, sitting with no substrate or hiding places and lighted by a 100-watt spotlight. That these scorpions, which love the dark, were still alive is a sign of their hardiness, but many may have died a day or two later.

Take a close look at the way the scorpion is offered. Crowded conditions may be fine temporarily; but even then, the scorpions should have access to shelter and water. Hot lights over the cage are a sign the dealer probably knows little about scorpions. Many crickets or mealworms roaming the cage without being eaten can be a sign that the scorpions are so far gone they may never eat again. Of course, even a temporary container should be clean.

Scorpions don't ask much, but any dealer offering scorpions must at least give them the proper temperature and humidity for the species, as well as proper shelter. Food is not essential while a scorpion is being offered for sale, as a healthy scorpion can go two or three weeks without feeding (longer in nature). Some dealers realize a customer will want to watch a potential purchase feed to ensure the animal is healthy. Withholding food from a hungry scorpion for a short time increases the likelihood that it will feed for the buyer.

Labels and Legalities

Never buy an unidentified scorpion! This could lead to serious problems should the scorpion prove to be a dangerous species. Currently, identification of scorpions is difficult, and not all are identifiable without being sent to one of the few experts on scorpions. There also is a strong tendency to sell small unidentified scorpions under nonsense names, such as "small yellow desert scorpion" or "black Asian scorpion." Not all keepers agree on the proper common names for widely available species, but there are some standards, and all suitable pet scorpions have valid scientific names as well as common names. Just because a scorpion is small and plain does not mean it is harmless—in fact, more often than not, "small and plain" is a good general description of many dangerously venomous species. Insist on a good scientific name—or at least a common name that can be researched and matched with a properly identified photo.

In some countries, governments regulate the sale and ownership of dangerous animals, which certainly includes scorpions. Even in areas without federal restrictions or with relatively mild limitations, there are many states and cities that enforce local laws that control or restrict the importation and keeping of dangerous animals. New York and Florida in the United States, for instance, strictly regulate local sales of all arachnids. In Europe, you may have to obtain a permit from both the federal and local governments and comply with strict, safe housing conditions. Because scorpion antivenins are not stocked in areas outside the natural range of the dangerous scorpion species, an accidental sting could be deadly, with no readily available treatment. Your dealer may require that you sign papers releasing the company from damage suits and guaranteeing you are of legal age (which varies from place to place).

Self-Collecting

How about picking up a few scorpions the next time you take a vacation? In the United States, this actually is possible if you live in or visit the southwestern states, where several dozen scorpion species in a variety of genera are found. (Only four species of scorpions are found in the eastern United States, three of these virtually restricted to Florida.) Unfortunately, you will find identification to be almost impossible and also will run the risk of coming into contact with the fairly common, dangerously venomous species, *Centruroides exilicauda*, which occurs in western New Mexico and in Arizona. Most of the scorpions in this area are harmless, however, though few adapt well to captivity and therefore are considered poor pets. Generally, giant hairy scorpions (genus *Hadrurus*) are the most widely desired species in the area.

In the rest of the world, collecting your own scorpions is trickier. Not only do you face local restrictions on collecting animals and problems with importing them back into your country, scorpions from tropical areas are also more likely to be dangerous. Most deadly scorpions come from the tropics of the Americas, Africa, and Asia, where field guides for identification of scorpions are rare and where dangerous species often look just like harmless ones. *Caution is advised.*

Actually, collecting scorpions is not difficult if you are willing to go out on a moonless night and hunt for them with an ultraviolet light (black light). For reasons that are not yet fully understood, the cuticle of all known scorpions contains chemicals that react to ultraviolet light by fluoresc-

A portable black light helped a hobbyist see this emperor scorpion wandering in a field at night. Chemicals in scorpions' bodies react with the black light, causing the animals to "glow" in the dark.

ing in shades of greenish yellow that stand out in sharp contrast to the background. When scorpiologists started using ultraviolet light to collect scorpions in the 1960s and 1970s, formerly rare species became common in collections, and many new species were found even in areas that were thought to be well-collected.

Low-cost, battery-operated, ultraviolet-light lamps are widely available at many hardware, camping, and big-box discount stores. The larger the light and the stronger its power, the farther the rays penetrate, and thus the more likely you are to see scorpions at longer distances.

Remember that ultraviolet light will not allow you to see snakes, spiders, cacti, and poison ivy, so be sure that you also carry a dependable headlight or flashlight and take care to follow paths and not reach into areas that might house snakes.

Scorpions also are collected by overturning leaf and bark litter during the day or night, and occasionally by digging out burrows. In areas where *Centruroides* bark scorpions are present, be aware that they often cling to the underside of rocks and bark and thus may be on the object you are lifting rather than on the ground. Many collectors have been stung by bark scorpions that were not noticed hanging on to a piece of lifted bark.

Scientific and commercial collectors often use pitfall traps—large tin cans or plastic buckets sunk into the soil or sand to the point that their lips are flush with the ground surface. The containers are covered with rocks or flat pieces of wood to protect the trapped scorpions from the sun, and a few punched holes in the bottom for drainage help prevent animals drowning during rains. Rows of dozens of pitfall traps are placed in the habitat, and the traps are checked at least once daily to prevent losses from heat, cannibalism, and predators. Some scorpions may be common in an area but never seem to enter pitfall traps, whereas others are abundantly trapped. Be aware that snakes and other dangerous animals also end up in pitfall traps.

Always check local laws to make sure it is legal to be wandering the countryside at night with an ultraviolet light—in some areas you cannot collect near highways and certainly not on private property or in parks. In many countries, night collecting may be unwise. Areas near the U.S.-Mexican border, for example, are regularly patrolled in an effort to regulate immigration and illegal drugs from Mexico into the United States, and a night of innocent scorpion hunting could easily be mistaken for illegal border-crossing or drug-running.

All Scorpions are Venomous

Hobbyists, especially beginners, must remember that scorpions are dangerous animals and that a few can be deadly—this is no joke! To be safe, you must either be able to identify scorpions accurately (which is not easy, especially for a beginner) or trust your dealer to have accurately labeled any scorpions offered for sale.

The sting of most scorpions is painful, producing local swelling and sometimes a tingling feeling or weakness near the sting site. The pain subsides after a day or two, with no long-lasting effects. All scorpions, even emperor scorpions (*Pandinus* species), can sting; some sting only when cornered and unable to retreat, whereas others are more defensive and sting whenever they get the chance. A single scorpion may sting repeatedly if trapped in clothing or in a shoe.

Not all the species of a genus have the same venom toxicity; so, from two species that are nearly identical in all other means, one may produce venom that is virtually harmless to humans, and the venom from the other is extremely dangerous. There even are reports that a single species may vary greatly in its venom effects over its geographical range. For example, the brown bark scorpion (*Centruroides gracilis*) is usually harmless if the specimen comes from Florida, but specimens imported from parts of Central America or the Caribbean are more dangerous.

Allergic Reactions

Keep in mind that the sting of even a harmless scorpion, such as the striped bark scorpion, *Centruroides vittatus*, may have different effects on different people. I've been stung several times by this species while collecting and, other than some cursing and hopping around, have suffered only minor, temporary pain. This also is the usual case for almost anyone stung by this species. Recently, however, there have been reports of more serious reactions to stings by striped bark scorpions, perhaps as a result of allergic reactions.

It long has been known that some people are more likely to suffer dangerous reactions to insect stings than others are; people have in fact died from as little as one or two honeybee stings. The same seems to be the case for scorpions, where a sting from a harmless species to a susceptible person can cause anaphylactic shock, leading to an irregular heartbeat, difficulty breathing, and sometimes death. There also is a possibility that being stung more than once by a harmless scorpion can lead to a serious, unexpected reaction at some time in the future if you are stung again. This happens with some plant toxins, such as poison ivy, where you may be able to walk through stands of the plant for years with no reaction and then suddenly break out when barely in contact with the plant.

The best way to get around possible allergic reactions is to never be stung. If you have other allergies and suspect you might be overly allergic to scorpion stings, consult your doctor for tests before buying a scorpion. There is no evidence that allergies to dogs, cats, or dust mites, for instance, have anything to do with allergies to scorpion stings.

The Rules

If you follow these simple rules and apply some common sense, you can select scorpions that can be safely kept as pets and not get stung.

- **Rule one** of scorpion selection and care is to never handle a scorpion with your hands or in a way that might allow escapes. We'll discuss ways of safely handling and inspecting scorpions shortly.
- **Rule two** is that scorpions with very large pincers (the palm is greatly swollen or inflated) generally are safe or at least not very dangerous. This includes the commonly available species of emperor and forest scorpions, genera *Pandinus* and *Heterometrus*, as well other members of the family Scorpionidae. The reasoning here is that scorpions with strong pincers on the pedipalps rely more on them to catch and tear apart food items than they rely on their venom. Many emperor scorpions seldom use their sting when subduing prey, but grab it and chop it to bits with the pincers and chelicerae. On the downside, large scorpions with strong pincers can produce a powerful pinch and their chelicerae a painful bite, both of which can draw blood; if this happens, clean the wound with alcohol and apply antibiotic ointment to prevent possible infection. Fortunately, there is no venom in the bite or pinch of a scorpion.
- **Rule three** is that any scorpion with slender pedipalp pincers is potentially dangerous, depending on their venom and sting to subdue prey. Though this rule tars many harmless scorpions with a broad brush, for a beginner it is the safest rule. Here the rationale is that the deadly scorpions all belong to the family Buthidae, a group characterized in part by slender pincers with long, thin fingers and a barely inflated palm. Not all buthids are dangerous, but because it is so difficult to identify genera and species in this very large family (eighty genera), it is best to err on the side of caution. Unfortunately, rule three would keep beginners from purchasing perfectly good pet scorpions such as giant

hairy scorpions (*Hadrurus*, family Caraboctonidae), a few decent vaejovids (*Vaejovis* and allies, family Vaejovidae), and even some bark scorpions (such as *Centruroides gracilis*, *C. hentzi*, and *C. vittatus*) that have proven to make decent pets despite belonging to a genus that includes deadly species. This is why you have to be able to rely on your dealer for accurate identifications and suggestions about keepable species.

Truly Dangerous Scorpions

It cannot be stated strongly enough: **There actually are a few species of scorpions with a sting that potentially could kill you, and some of these species are sold in the pet trade.**

Most scorpions that are considered deadly produce a venom that acts as a neurotoxin and causes widespread damage to the body, including convulsions, irregular heartbeat, and respiratory paralysis. In many cases, there is little or no pain at the site of the sting, and the first symptoms may not appear for hours after the stinging. Death can result in as little as three or four hours, or in as long as two or three days. The venom is exceedingly toxic and is especially dangerous for children under three or four years old (because of their small body size), the aged, and those with autoimmune diseases, though, some scorpions are capable of killing a healthy adult human as well.

Though the venom of the striped bark scorpion (*Centruroides vittatus*) is not as toxic as that of its relatives in the dangerous family Buthidae, it still should be avoided, especially by beginning hobbyists.

Treatment for stings from dangerous scorpions is supportive (for instance, respiratory assistance, IV glucose, constant monitoring, and control of blood electrolytes) in order to reduce the potential for respiratory paralysis and heart failure. Generally, an antivenin produced strictly to treat the sting from a particular scorpion species is administered. However, few hospitals outside the range of dangerous scorpions actually stock antivenin; and the antivenin itself may cause serious—sometimes fatal—allergic reactions if it is based on horse serum, to which many people are allergic. The effectiveness of first aid is uncertain, though cooling the sting site with ice may slow the spread of the venom.

Trying to extract the venom by sucking or using a small plunger pump also is of uncertain value as such measures would have to be applied within seconds of the sting, which may be impossible.

Genera to Avoid

The twenty to twenty-five species that have proven to be consistently dangerous belong to the following genera, all members of the family Buthidae:

- *Androctonus* (northern Africa to India)
- *Centruroides* (southern United States, Caribbean, and tropical Americas)
- *Hottenotta* (Africa to India)
- *Leiurus* (northern Africa and Middle East)
- *Mesobuthus* (Middle East to China)
- *Parabuthus* (Africa and Middle East)
- *Tityus* (Caribbean and tropical Americas)

Some species of *Androctonus*, *Leiurus*, and *Tityus* seem to be especially toxic and should be avoided by all keepers. Species of several other buthid genera not on this list occasionally have been accused of producing serious—even deadly—stings, but these may be the result of individual allergic reactions. Beginners should be leery of all buthid scorpions.

CHAPTER 3

HOUSING AND FEEDING

Most scorpions are easy to keep and will thrive in a small vivarium with minimal furnishings. Some attention to temperature and humidity extremes is essential, however, and not all scorpions can be kept in the same type of vivarium. First though, let's discuss how to move your scorpion around without being stung.

Moving Scorpions Safely

The temptation to pick up and display a large, placid scorpion such as an emperor scorpion (*Pandinus* spp.) can be strong, but the best rule is to never touch a scorpion and never keep it outside its cage for more than the few moments necessary to move it from one container to another. All scorpions will sting, sometimes without obvious provocation, and the sting of all species is troubling even if there are no problems of toxicity or allergic reactions.

The best way to move a scorpion around is to gently direct it into a clear plastic jar or box by using a rod or a pair

A small, simple, clear plastic critter box makes a suitable home for your pet scorpion.

of long forceps. As soon as the scorpion backs into the container, cover it securely with a lid that cannot fall off or be pushed off by the scorpion (scorpions can be very strong and persistent). Watch for the scorpion moving forward rather than backward as you try to get it into the container—many bark scorpions (genus *Centruroides*) will quickly try to climb up the wooden shaft of a pencil or dowel rod toward your hand, which can be dangerous. A glass rod, such as is sold to stir chemicals, is much safer, as scorpions cannot climb unpitted glass. The container must have a fairly wide mouth, as most scorpions hold the pedipalps widely extended when they are challenged. Using a clear container makes it easier to be sure the scorpion is still in the jar or box and has not escaped.

While the scorpion is in the container and securely capped, take the opportunity to give it a close look and perhaps see if you can determine its sex by examining the pectines or by looking with a magnifying glass at the genital operculum. If you are trying to identify a scorpion, this is your chance to look at the crests on the pedipalps, body, and tail in detail and even to notice the shape of the spiracles and how the spines of the legs are arranged. Never leave a scorpion in a transfer container for very long, as it could overheat.

Be very careful when the scorpion leaves the transfer container that it does not double back toward your hand. Use a rod to make sure it goes in the right direction.

If you are more adventuresome and want to take a chance with a safe species, you can use long forceps (the 10-inch, 25-cm, size is fine) to pick up a scorpion by gently but firmly grasping its tail below the telson. Scorpions with large bodies and slender tails should not be picked up this way, as the weight of the body could damage the tail. To reduce the chance of damage to the scorpion's tail caused by squeezing, coat about 3 inches (7.6 cm) of the forceps tips with rubber. Liquid rubber is available at hardware stores; dip the forceps into the liquid, let the coating dry, and apply at least two more coats. You also can try to cover the tips with plastic tubing of the appropriate diameter. Putting plastic tubing on the forceps is not something I favor, as I always end up applying more pressure to make up for the smoothness of the tubing; sandpapering the tubing may help.

Basic Vivaria

Your scorpion is a pet, and you want it to thrive in its new home. There are important aspects of safety—for you and for the scorpion—to be considered when setting up a scorpion vivarium.

The Container

Scorpions neither need nor want large containers, and they have less trouble finding food and cover in a small vivarium. A container roughly four or five times the length of the scorpion will be adequate. For small scorpions less than 3 inches (7.6 cm) long, a 1-gallon (3.8-L) container will provide plenty of room, whereas the largest *Pandinus* and *Heterometrus* species can live their entire lives in a 2-gallon (7.6-L) vivarium.

Today, the easiest to purchase and probably most suitable containers for scorpions are the clear plastic critter boxes that have slightly slanted sides and slotted tops. The plastic is dense enough that most scorpions (except perhaps some of the smallest bark scorpions) cannot climb the sides unless you allow dirt and calcium deposits to accumulate, which give the scorpion a leg up. Two convenient sizes are the 1-quart (0.9-L), which is about 6 inches long × 3 inches wide × 3 inches high (15 × 7.6 × 7.6 cm), and the approximate 2-gallon (7.6-L), which is roughly 10 inches long × 6 inches wide × 6 inches high (25 × 15 × 15 cm). The smaller size can hold a single, small scorpion specimen, and the larger size will be roomy enough for a single, large emperor or forest scorpion (though the height may be a bit low). Make sure that the height of the container is more than the combined length of the scorpion plus the depth of its substrate so that the scorpion cannot reach the top edges of the container when standing.

Unfortunately, the slotted lids of critter boxes are not only almost useless, they actually can be dangerous. The slots are more than wide enough for many small scorpions to walk through with ease if the container should be accidentally turned over, and the number of slots will quickly allow a substrate to dry up. Additionally, these lids are fragile and easily broken by a short drop onto a hard surface. I suggest using the lid just as a securing device under which you can wedge a sheet of aluminum screen mesh and/or plastic. Cut the mesh (window screen is widely available in hardware stores) about 1.5 inches (3.8 cm) wider than the top edge of the critter box and place it evenly over the top of the container, firmly pushing the slotted lid into position and thus forming a mesh lid under the slotted plastic top. Plastic sheeting (heavy transparent plastic sold to insulate windows is great for this) placed below the screening and held in place by the lid helps increase humidity in the container, but it should not cover more than half the top; make sure it is cut sufficiently wide to safely overlap the top of the plastic container.

Other types of plastic containers can be used, too; some have the advantage of being tougher and less subject to breaking if dropped. Small holes can be drilled into the upper sides and lid of solid plastic containers by using a hot nail, a fine drill or punch, or a soldering iron with a narrow tip. In all cases, make sure the lid covers the frame completely and tightly. A scorpion can slip through a very narrow opening and escape if the lid leaves a space next to the top of the container at any point.

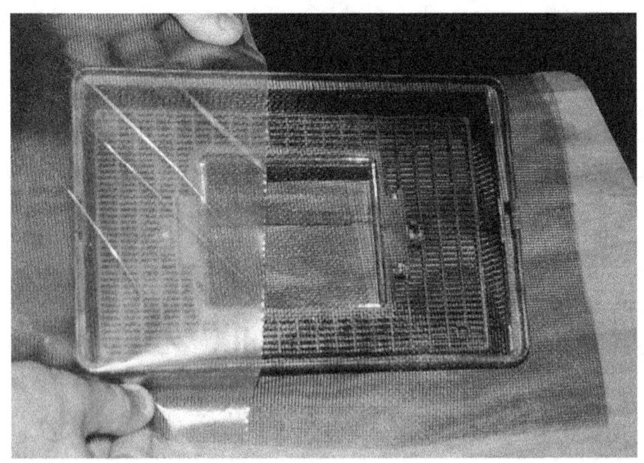

Screen mesh placed under the critter box's lid helps prevent escapes by small specimens, which can slip through the lid's slots. Plastic sheeting over half of the lid will increase humidity in the vivarium.

Glass containers such as 1-gallon (3.8-L) or 5-gallon (19-L) aquariums can be used for scorpions, but they are relatively expensive, heavy, breakable, and seldom have lids that fit securely at all sides. Additionally, most small scorpions can climb up the silicone sealer used to hold the glass sides of an aquarium together, which is basically an invitation to explore the underside of the lid—never a safe situation. The smaller and lighter the scorpion, the more likely it is to climb corners; some very small scorpions can even wedge their way up corners just using the tension from water droplets. No scorpion can climb clean, unpitted glass nor most smooth, hard plastics—but remember, it is difficult to keep glass and plastic clean and unpitted.

All vivaria containing scorpions should be carefully labeled with information on the name of the species, its danger level, and when and where it was purchased or collected.

Double-Caging

If you have any potentially dangerous scorpions, you must keep them in double cages in order to limit the possibilities of escapes. Scorpions, especially the smaller ones, are wonderful escape artists, able to squeeze through small cracks and, in the case of bark scorpions and their allies, rapidly climb to freedom if their legs can gain any purchase on the sides of the vivarium. The most harmful scorpions are small enough to be good at making escapes, which can lead to dangerous situations; scorpions head for tight, dark areas when they escape and probably will never be found during a search.

Reduce the possibility of escapes by double-caging, putting all cages with dangerous scorpions within a larger plastic container such as a deep, plastic storage box. Make sure the lid can be securely fastened and, preferably, locked. To allow air movement and to prevent humidity from building up, punch, drill, or burn quarter-inch (6.4-mm) holes in several series around the top of the safety cage and in several places in the lid. Sometimes it is sufficient to place a small critter box within a larger one, making sure that there is strong mesh over the tops of both boxes.

All cages housing dangerous scorpions must be fully labeled. Never allow children near these cages, and always handle scorpions and the cages that hold them with great care.

For added security, keep your scorpion's vivarium inside a larger container.

Substrates

Arguments about the best substrates for scorpions could go on for pages, but this really is not necessary. Scorpions are burrowers in sandy soils and in moist soils, or they hide under bark and rocks on the forest or savanna floor; only a few are found under bark on standing trees or wedged into rock crevices. The universal substrate for scorpions is a sand and potting soil mix, varying the ratio of ingredients to accord with the natural habitat of the scorpion. A desert burrower such as a giant hairy scorpion (genus *Hadrurus*) likes a mix high on the sand end, with just enough soil to hold the sand together for burrowing. Try a mix of 75 percent sand and 25 percent soil and see how your scorpion likes it. Emperor scorpions and forest scorpions (genera *Pandinus* and *Heterometrus*) come from forest and savanna areas, which have more moisture, and accordingly like a mix heavy on soil and light on sand. For these scorpions, try 75 percent soil and 25 percent sand. Bark scorpions (*Centruroides*) rest under bark and rocks and don't normally burrow; keep them on a 50:50 mix of sand and soil with some flat pieces of bark for cover and they will be happy.

Watch the behavior of your scorpion to help determine the proper substrate mix. If too much sand is in the mix, burrows will crumble easily. You really don't have to become hi-tech to give your scorpion a good base in the vivarium. Burrowers need a deeper base than do nonburrowers, but remember that the scorpion must never be able to reach the rim of the cage while standing upright on the substrate. More than 3 inches (7.6 cm) of substrate is unnecessary for any scorpion, and it will allow the burrowers to disappear in the bottom and never be seen.

Cage Accessories

Plants, ceramic structures, and other decorations are not necessary in a scorpion vivarium; they will not be appreciated, and they could allow a scorpion to find a way to reach the top and escape or to hide and attack while you are cleaning. Almost all scorpions will use a bit of cork bark or a small rock as shelter for their burrow or hiding spot. Commonly, you start a scorpion burrow by using a pencil to make a small hole under a rock or piece of bark that is partially embedded in the substrate. The scorpion will use this preexisting entrance and enlarge it to fit.

The only cage accessory your scorpion really needs is a small water bowl—about 3 inches (7.6 cm) in diameter (less for a small scorpion) and an inch (2.5 cm) or less in depth. Most scorpions will drink when water is provided, holding the mouth over the water and sucking it in or using the chelicerae and pedipalps to help direct water into the mouth. If you have small scorpions, put a few pebbles on the bottom of the water bowl so they do not

A small piece of cork bark (left) atop moistened potting soil (right) is all the decoration you need to include in your scorpion's vivarium.

accidentally drown—some scorpions actually will soak in shallow water on occasion.

True desert scorpions seldom drink, so they do not need a water bowl. Try putting a water bowl in their vivarium for one day per week and see if the scorpion uses it or not.

Yes, a vivarium without elaborate decorations will look empty, but this actually makes cleaning easier and safer. The scorpion will not mind at all.

Temperature and Humidity

Most scorpions do well at just a bit warmer than normal room temperature and in moderately moist air, and they are not especially sensitive to small deviations.

Temperature

No scorpion likes very warm temperatures. Heat can kill a scorpion if the temperature isn't allowed to at least drop at night. Even desert scorpions avoid the sun and hot sands, coming out to hunt and mate at night after the temperature has cooled.

Most scorpions can be kept at moderately warm room temperatures. Try to maintain a range of roughly 75°F to 86°F (24°C to 30°C) during the day, dropping a few degrees at night. The simplest way to achieve these temperatures is to use a red heat lamp (scorpions don't see red and aren't bothered by it) positioned over the vivarium at an appropriate distance. Use an accurate thermometer on the floor of the vivarium to make sure you know what the real temperature is in the cage, not just the air temperature in the room. Simple and inexpensive electronic thermometers with sensors at the end of long cords are widely available in electronics stores and some discount stores. Heat lamps are inexpensive and can be found in many sizes and designs at any pet shop (check the reptile section). Additionally, these lamps can be placed in clip-on aluminum domes to help concentrate their heat. Use a heavy-duty appliance timer to turn heat lamps on and off, giving the scorpions a warm day of about fourteen hours and a cooler night of ten hours.

I do not recommend undertank heating, either with pads or heat strips, for scorpions because this makes substrate warmer than the air, which is the opposite of any natural habitat. Scorpions burrow to avoid heat from above; the farther they dig into the soil, the cooler it is. Some hobbyists have used heat strips along the backs of their cages, anchored to vertical surfaces in a shelving unit. This may work, but it seems best to me to have heat only from above. Heat strips have caused fires when improperly installed, and both heating pads and heat strips may melt plastic vivaria. Caution is advised.

Importance of Daylight

Heat lamps do not replace natural daylight, which scorpions need to control their biological rhythms such as molting and mating. If at all possible, you should house scorpions in an area where a window allows them to sense the passing of days and seasons. If you store scorpions in a dark closet (as often is the case), then at least place a fluorescent tube across the top of the closet and connect it to a timer that you can set to simulate natural day lengths. There is little evidence that scorpions from tropical countries have to adapt to light cycles in the temperate zone to behave normally.

Humidity

The humidity of the substrate, which is reflected in the relative humidity of the vivarium, should not drop below roughly 50 percent, and most species will tolerate 60 to 70 percent with no problems.

Though desert scorpions do not thrive in very humid conditions, this does not mean that they need dry vivaria. Their burrows often are relatively humid compared with the surface, as you might expect. It seems their burrows should also be kept at 50 percent relative humidity even if the surface drops to 30 or 40 percent (not recommended). Some hobbyists have devised ways of assuring the burrow stays humid while the surface stays dry. One way is to place a thick layer of aquarium gravel or plaster of Paris over the bottom under the substrate to hold humidity. Another way is to place small vertical pipes (PVC or even drinking straws) in substrate at the corners and add water to these on a weekly basis, which keeps the surface dry. There really is little evidence, however, that such complicated setups are necessary for scorpions commonly kept in vivaria.

Forest scorpions and most bark scorpions seem to like higher humidity near the surface where they normally spend the day. Forest scorpions, as you might expect, can live well when kept above 70 percent humidity, but they really aren't choosy.

Many scorpions, including bark scorpions and forest scorpions, appreciate having their container lightly misted with lukewarm water once a week (more often for forest scorpions). Desert scorpions seldom need their container misted, just as they seldom need drinking water. Never overdo misting—too much surface water can lead to bacterial growths. And never directly mist a scorpion—a wet scorpion can really get angry!

Basic Foods for Scorpions

Interestingly, we know almost nothing about the nutritional requirements of scorpions. We *do* know that all species are carnivores, feeding on almost any soft-bodied invertebrates and small vertebrates they can catch. Their prey contain a variety of vitamins and minerals, which in turn are absorbed by the scorpion. A few hours before offering to a scorpion, prey insects can be fed a mix of shredded carrots and green lettuce, often with a bit of calcium supplement powder added; this is called gut-loading, and is a good way of ensuring your scorpion eats nutrient-rich food.

Many hobbyists have maintained large numbers of scorpions by feeding them cultured crickets (*Acheta domesticus*), mealworm larvae (*Tenebrio molitor*), and superworm larvae (*Zophobas morio*). These three insects are bred in gigantic numbers, are readily available year-round at reasonable prices, and can be housed and gut-loaded in small containers while waiting to become scorpion food. Crickets and mealworms are available in a variety of sizes to suit even small scorpions. You don't have to culture the bugs yourself unless you have a large scorpion collection, because individual scorpions need only a few food items a week.

Scorpions chew and rip apart their prey, swallowing mostly body juices. This cricket will soon be just a crumpled shell.

Even if offering cultured bugs, you may supplement the diet with wild-caught spiders and insects such as grasshoppers and dragonflies (but never from areas that have been sprayed with pesticides or from roadsides, where pollution is greater). Most scorpions avoid hard insects (such as beetles), stinging wasps, and ants. (Though ants will be eaten on occasion, they are more likely to become pests that prey on scorpions.) Scorpions will also eat small frogs and lizards, though I hate to think of these vertebrates becoming prey. Feeding newborn mice to large emperor scorpions leads to a bloody mess, but many large scorpions do like the occasional mouse as a "treat."

Scorpions are notorious for undergoing long fasts and then eating until their bodies literally expand before your eyes. Such behaviors probably are related to natural cycles where prey animals only become available for a few weeks or months during the year, so the scorpion must eat all it can, while it can, and just wait out the year for the next feeding. This is totally unnecessary in the vivarium, however, where you can feed a scorpion on a regular schedule.

Small, growing scorpions generally are given three or four food items (small crickets or mealworms) two or three times a week. Larger scorpions may take as many as six or eight crickets or two or three superworms once a week. Live foods are preferred, though sometimes small scorpions will take crushed crickets with body fluids oozing out. Using forceps, remove uneaten food and prey remains the same day it is fed.

CHAPTER 4

BREEDING

Although imported, gravid, female scorpions often give birth in the vivarium, few keepers have had much luck producing a captive-bred second generation. Surprisingly, even the most abundant species, such as the emperor scorpion, are virtually all imported from their native countries. There they may be farmed—kept in large pens or containers and allowed to mate freely in natural conditions. The young are harvested when old enough to survive shipment to the purchaser. This is not considered to be true captive breeding, where a keeper puts together the parents at a specific time and allows them to mate; the keeper then cares for the mother until she gives birth, later raising the young until they are mature or nearly so.

One group of scorpions can be bred in captivity with some consistency: the bark scorpions (*Centruroides* species). Unfortunately, these are considered dangerous and are prone to escape, as well as being abundant in the wild and easily collected, so few keepers put much effort into

It won't be much longer before these baby desert scorpions (*Centruroides* sp.) leave their mother's back. Young scorpions live about the first two weeks of life atop the mother.

breeding them. If several pairs are placed together in a large vivarium, fed, and kept at the proper temperature and humidity, you are almost assured of successful breeding. Additionally, some keepers have had moderate success breeding emperor scorpions (genus *Pandinus*), dangerously toxic fat-tailed scorpions (genus *Androctonus*), as well as *Hadruroides* species from South America and a scattering of other species.

Sexing Scorpions

Except for one species of *Tityus* known to be parthenogenetic—females reproduce without mating with males (meaning that males have never been found)—and reports that a few other scorpion species might be parthenogenetic in some populations, all scorpions have two distinct sexes.

In many common species it is possible to distinguish sexes by comparing the appearance of adults among a large group. As a rule, male scorpions tend to be more slender and smaller than females, and they sometimes have elongated ring segments forming the tail or postabdomen. In some cases, such as flat rock scorpions (*Hadogenes troglodytes*), death stalkers (*Leiurus quinquestriatus*), and many bark scorpions (*Centruroides* species), the tail of the male is distinctly thinner and sometimes as much as a third longer than the female's, so the sexes are easy to distinguish if both are in the same container. Male scorpions generally (but not always) have more slender pedipalp pincers than females have, the fingers also being longer. Color differences between scorpion sexes seldom are reported.

If a scorpion is placed in a clear plastic or glass container and turned so you can check the underside, you will see the pectines (see also chapter 1). In almost all common scorpions the pectines of the male are more developed than in the female. Within a species, a male tends to have longer pecten bases and more teeth that usually are more slender than in females. This is relative, of course, so you usually need both sexes to be able to compare. Tooth counts are not absolute, as they nearly always overlap a bit (the female may have nineteen to twenty-six teeth, the male twenty-four to thirty, for instance), but they give you a chance to make an educated guess.

In a very large scorpion or if magnification is available, you can check the genital operculum for further clues. In males, the two plates of the operculum may be distinctly separated down the midline, exposing a pair of short but sometimes obvious fleshy projections that are the paired lobes of the penis. In females, the two sections often are fused together, and no projections are visible. Male scorpions of all common species (except for giant hairy scorpions, genus *Hadrurus*) have visible penises.

Few details on sexing scorpions by differences in genital operculum have been published. But, with experience and several specimens at your disposal, you can make a good guess at the sex of almost all common scorpions (though even the experts make mistakes).

Mating

Mating in scorpions may follow a short period of relatively cool and dry weather, when the scorpion feeds little and seldom leaves its burrow or hiding spot. Some scorpions, however, seem to breed year-round. If well-fed, adult, and healthy, it is likely that putting two scorpions of different sexes together will lead to mating behavior.

Be aware that mating can be very rough on both animals. Males sometimes sting females before mating to calm them, which can eventually lead to death. Females often attack males after mating and kill and eat them. Always carefully watch a pair of scorpions when put together, and make sure the male has lots of safe retreats. Attempting to separate two enraged scorpions may be dangerous, as their actions become unpredictable.

When the male detects the presence of the female (the male usually is put in the female's vivarium), he begins to court her by thumping his legs and sometimes tail on the substrate. His pectines apparently can detect chemicals emitted by the female that tell him if she is willing to mate. If she shows interest in his activities, usually by doing some thumping of her own and approaching him, he grabs her pedipalp pincers in his and begins a circling dance, leading the female across the floor of the vivarium. They also may interlock the chelicerae, which helps assure the male that he will not be attacked.

The purpose of the dance is thought to be a search for a hard, dry, smooth surface, such as a flat rock. This is essential because male scorpions produce sperm held in a slender tube, the spermatophore, that becomes glued to the substrate at one end and has a spring mechanism at the other. The male projects a spermatophore from his genital operculum onto the proper hard surface, makes sure it is glued correctly so it is upright, and then waltzes the female over the spermatophore. The animals jiggle around until the spermatophore is perfectly centered under the female and the spring mechanism hooks into her genital operculum. Once the spring is triggered, it projects the sperm packet into her body, where it can be stored for several months. At this point, mating technically is over, but the male still has to make his safe escape. The two animals release the grip on their pedipalps and chelicerae and back away from each other. The male then usually makes a sprint for cover, while the female retreats and ignores him—or follows and attacks him. As I said, mating is rough in scorpions.

Mother's Work

Female scorpions have a complex system of tubes known as the ovariuterus in which the eggs are fertilized and develop into large embryos before being born. In some scorpions, the embryos are nourished on large amounts of yolk; but in a few species, there is a membrane that passes nutrients from the mother's body to the developing embryo, and yolk is virtually absent.

Fertilization of the eggs probably takes place several weeks after mating. A gravid female slowly fills with developing young, which often can be seen through the thin membranes on the sides of her body. Just before giving birth, a female may appear to be about to burst with white young. The embryos take anywhere from three months to eight months, sometimes longer, to develop fully. During

Though eating her own babies is rare for a scorpion mother, it does sometimes occur. This emperor scorpion mother has plucked one of her babies off her back to make herself a meal.

this time the mother may be more secretive than usual, may drink more, and during the final month may not eat. Don't bother the female, and don't treat her any differently than usual.

Birth usually occurs under cover or in a burrow and seldom is seen. Generally, the mother delivers all her young in less than an hour. Each white baby pops out of the genital operculum and falls toward the ground. The mother usually assumes a position where the front legs are held together under the body to form a "birth basket" that may gather the babies as they are born and hold them until they shed a thin birth membrane (not present in all scorpions). As soon as the soft white young—which have distinct tails and legs but move with difficulty—are free, they crawl up one of the mother's legs and reach her back. The entire litter of young eventually will gather on the mother's back, where they will stay, moving little and not feeding, for ten days to a month.

These baby scorpions are born in what is termed the first instar. (An instar is a period of time between molts in invertebrates; the word also is used to indicate the age of the animals.) First instar young are poorly developed and either are still absorbing their yolk (if present), or are using other stored nutrients to complete their development. Eventually, usually in less than a month, they will molt (all at about the same time) and become second instars. Second instars have dark coloration (like the parent), have the legs and tail fully developed and usable, and are quite capable of stinging prey and tearing it apart. After the second instars harden their exoskeletons and all their internal organs become fully functional, they climb or jump down from their mother's back and begin independent lives. At this point, they may begin to prey on each other, so it is time to move each little scorpion into its own container.

The number of young in a litter of scorpions varies from about a dozen in some very large species to more than one hundred in others. The general rule of thumb is: Large scorpions produce a small number of large-sized young; small scorpions produce a greater number of small-sized young; and the smallest scorpions may only produce half a dozen young, as there is not enough room in the mother's body for more embryos than that. As extremes among common species, *Pandinus imperator* gives birth to roughly ten to twenty-five young, whereas *Centruroides gracilis* may have fifty to nearly one hundred young.

Caring for Baby Scorpions

Second instar scorpions are just like tiny versions of their parents, hiding or digging small burrows and feeding on small insects. They usually feed on small crickets but can also take flightless fruit flies (*Drosophila* species) and micromealworms (*Tenebrio obscurus*). Some baby scorpions will take freshly killed and crushed food insects, but

A first instar desert scorpion (*Centruroides* sp.) measures only about one-fourth the size of a dime.

Two young Texas bark scorpions crawl out from under a rock, the one on top showing pale coloration, evidence of a recent molt.

most prefer to catch their own. Remember that their cage has to be small or they will not be able to find their food.

Young scorpions lose water more quickly than do adults, so be sure that the humidity in their vivarium and burrow is at least 60 percent (a bit higher in the actual burrow). Some will take water from tiny containers such as bottle caps; they drown easily, however, so be sure to add a few pebbles to the dish.

Scorpions grow by molting their skins, but molting seems to be restricted to the period before they become sexually mature. Most scorpions undergo six to eight molts during the one to three years it takes to reach maturity.

CHAPTER 5

HEALTH

Maintaining suitable conditions of temperature and humidity as well as providing adequate food should keep your scorpion healthy. Most scorpions live three to five years in properly kept vivaria, with some larger species living a decade or more.

Environmental Problems

Extremes are harmful to scorpions, whether high or low temperatures or dry or moist conditions. Temperatures higher than roughly 95°F (35°C) are likely to be fatal for most scorpions, at least when they are not allowed to burrow into a cooler substrate and have access to water. Overheated scorpions may act disoriented, circle endlessly, and eventually collapse and die. Even the best-adapted desert scorpions—those living in areas where the daytime temperature may exceed 120°F (49°C) for weeks at a time—die at high temperatures. In nature, scorpions burrow during the day and come out at night when temperatures drop to suitable levels; they also can go long periods without feeding and may become totally inactive during peak summer temperatures.

Cold temperatures are less dangerous to scorpions, merely causing the animals to become inactive and to cease feeding. Most of the commonly kept scorpions remain active when the temperature drops below 70°F (21°C), and many smaller species are still going strong when the temperature drops to 60°F (16°C). Below this, scorpions generally become comatose and may eventually die from starvation and internal problems.

No scorpion should be kept in a totally dry vivarium, and it is likely that a relative humidity less than 40 percent may be harmful to most scorpions if they are not allowed to look for moist spots and provided water to drink. On the other hand, high relative humidity is harmful to most scorpions from dry habitats (such as the southwestern United States and the deserts of northern Africa and the Middle East) and should not exceed about 60 percent.

Forcing a scorpion to live in a brightly lit vivarium certainly will be stressful, and it may lead to an early death. Scorpions are nocturnal animals, though some hunt at dusk and dawn; they are seldom active in full daylight.

Additionally, avoid placing a scorpion where there is constant activity that causes vibrations in the substrate. Many sense organs on the legs of scorpions are devoted to sensing vibrations and air movements, and certainly such stress will eventually prove harmful. Scorpions lack true ears and, consequently, don't seem to notice airborne sounds; but remember that a loud television or radio in a room can cause vibrations in solid objects, including the vivarium.

Molting

By the time it is mature, a scorpion has ceased molting. It appears (from the limited information available) that scorpions molt six to eight times between being born and reaching adult size (which takes one to three years); after that, they no longer grow and thus cease molting. Very old scorpions commonly have scars where appendages were lost, and most of their setae and trichobothria on the legs and hands of the pedipalps may be abraded or lost entirely. Even the teeth of the chelicerae may be worn down.

During the molt, fluid accumulates between the old exoskeleton and the soft, new skin contained under the old one. Scorpions about to molt become relatively inactive and may stop eating for a few days, as the lining of the gut and the internal structures of the legs and muscles also are replaced during a molt. A molting scorpion hides deep within a burrow or in a protected spot because it will become extremely vulnerable to predators during and after the molt; during this time, its new skin is very soft and can be easily penetrated by an enemy. Molting usually takes place at night.

A break develops in the old exoskeleton along the front of the carapace above the chelicerae and then extends along the sides between the top and bottom of the cephalothorax. When the top front of the carapace pops open, the scorpion draws its body up through the opening and slowly pulls out the body and appendages, including the tail and telson with its sting. Scorpions seem to have few troubles with molts, and losses are few.

The newly molted scorpion is soft and weakly pigmented. It takes several days to a week or more to fully harden the exoskeleton as calcium salts in the cuticle react with the air to harden and darken the skin. During this time, the scorpion remains in hiding and may bury itself in the substrate. It does not feed during this period, and it seldom tries to move. Once the skin hardens (internally and externally), life continues as normal until the next molt, which may be two or three months later.

The molted scorpion skin is an exact representation of the scorpion, down to the position of its bristles and delicate internal structures in the legs. It sometimes even retains some of the coloration of the scorpion. Though shed skins are delicate, some hobbyists carefully dry them and then pin them in a box with insect pins. If you carefully label all shed skins with species, locality (if self-collected), and date of molting, you will have interesting mementos of your pet's growth.

Accidents

Though they are agile, tough animals, scorpions can injure themselves by falling from high spots, rubbing against sharp surfaces, or interacting with other animals.

Because of their light weight, smaller scorpions can fall great distances with little chance of harm, though there always is a chance of breaking the tail. Large, heavy scorpions are likely to be seriously damaged by falls of more than 2 to 3 feet (60 to 91 cm). In nature, few scorpions will jump, nor may they necessarily see (or understand) that a cage lip has an edge and then a dangerous drop. So, ensure that your scorpion's environment is safe to prevent falls.

Cuts are likely if you place a thorny branch or even some cacti in a small vivarium that does not allow the scorpion sufficient maneuvering room to avoid injury while chasing prey. Though the cuticle of a scorpion is thick and hard, remember that there is a relatively thin membrane between the different segments to allow movement. The membrane on the side of the body is wide and quite thin and thus can easily tear.

Because a scorpion's blood is not contained within distinct vessels but just moves slowly around the internal organs and appendages, any rupture of the cuticle will result in bleeding and possibly significant loss of blood before clotting occurs. If the cuticle is broken, allow the scorpion to drink freely to replace fluids, and *very carefully* attempt to place a small piece of tissue paper over the opening to speed clotting. Sometimes fine cornstarch sprinkled over the cut (*not* over the entire animal, which could clog the spiracles and suffocate it) will aid clotting as well. A wounded scorpion will not appreciate your attentions and may not even know that it is bleeding. Don't get stung while attempting first aid!

Scorpions generally seem to be immune to their own venom. (Accidentally stinging yourself seems to happen in the scorpion world.) Within a species, the venom may have little or no effect. However, when a male scorpion stings a female of the same species in one of her appendages during mating, that appendage often becomes inactive and then drops off; so, obviously, intraspecific immunity is not absolute. Scorpions of one species commonly sting and kill other species, and large specimens of a species may also sting smaller members of their own species, even if the venom does not cause death.

CHAPTER 6

FAMILIAR SCORPIONS

With more than fifteen hundred described species of scorpions, one might think that dozens of species should be available for the vivarium. The reality is that there probably are fewer than twenty-five scorpion species (or at least groups of virtually identical species) sold on a regular basis in the United States and Europe, excluding, of course, locally collected species of little interest to dealers. Perhaps a third of these are dangerously venomous and should not be kept by beginners—and probably should not be offered for public sale. Additionally, miscellaneous (often unidentified or misidentified) scorpions appear from time to time but seldom become established in the hobby. Beginners should never purchase unidentified scorpions or specimens about which there is any doubt as to their genus and family.

The scorpions described below are just a hint of those you might find by visiting a shop or expo specializing in invertebrate pets, or by checking through the Internet and magazine ads. (Beware: shipment of scorpions across state lines can be complicated and is often illegal.) Use these write-ups to get a taste for what species are available on the market today.

Relatively Safe Scorpions

The species or groups of related species you will read about in this section can be kept by the careful beginner or intermediate hobbyist. (Their venom is not known to be particularly dangerous, though stings can be extremely painful; remember that individual allergic reactions to any scorpion sting are possible.) These scorpions are listed by their general suitability for beginners.

Emperor Scorpion, *Pandinus imperator* Family Scorpionidae

This gigantic scorpion, often more than 6 inches (15 cm) long from the front of the carapace to the tip of the telson, remains the best scorpion for a beginner. A native of tropical forests and savanna edges over western Africa from Senegal to the Congo, it spends the day in shallow burrows under tree trunks, leaf litter, and trash where the soil is moist and rich, coming out at night to feed on native roaches, grasshoppers, frogs, and lizards.

The body is thick and heavy for a scorpion, with a narrow, weak tail, though the pedipalp pincers are gigantic (especially in large males) and are covered with many rows and clusters of large tubercles. The front margin of the carapace is deeply notched at the center. Overall color is a dark, deep brown to black with a faint greenish tinge when observed in the proper light. This species is protected to some extent, and many specimens on the market come from commercial farms in Togo and Benin. The emperor also is one of the most consistent breeders in the vivarium.

Keeping an emperor scorpion is simple. Give it plenty of room—at least 2 gallons (7.6 L)—in a vivarium deep enough to prevent the scorpion from reaching the top when standing. The substrate should be at least 3 inches (7.6 cm) of potting soil, preferably with some live peat moss added. Keep the humidity level between 75 percent and 85 percent,

One of the largest and most harmless scorpions, *Pandinus imperator* is the most suitable species for a beginning hobbyist.

and the temperature between 80°F and 85°F (26.5°C and 29.5°C), dropping only a small amount at night. Feed liberally on crickets, superworms, and other meaty foods, including the occasional newborn mouse as a "treat." Though you can keep several *Pandinus* together as a small colony, you are better off housing them separately.

Emperor scorpions are among the few scorpions that can be handled with relative safety, as they seldom sting and generally are calm. However, even a pet emperor can become frightened or annoyed and pinch (they can draw blood) or even sting. It's best to leave the handling to professionals.

Occasionally, other emperor scorpion species are offered for sale under a variety of names. Some are a bit more reddish in shade than the true emperor and may have small differences in the structure of the pedipalp pincers. But these species are difficult to identify correctly. These usually come from eastern Africa and can be treated much like *Pandinus imperator*. Some of these emperors may be relatively nervous and much more ready to sting than the West African emperor.

Forest Scorpions, *Heterometrus* Species Family Scorpionidae

These are the tropical Asian equivalents of the emperor scorpions and match them in most respects, including overall appearance and size. As a rule, though, they have fewer granules on the pincers, which appear smoother.

More than thirty species now are recognized in the genus, which ranges from India to Indonesia, with most species imported from Southeast Asia. The names placed on these scorpions by dealers (such as *Heterometrus longimanus* and *H. spinifer*) often are wrong because the species are almost impossible to identify from available literature; but fortunately, all commonly imported species from this family can be kept the same way and are much alike in temperament.

Keep forest scorpions much as you would emperor scorpions—in a large, tall vivarium with a base of potting soil, and feed them a variety of meaty foods. You can try to keep several specimens together in a very large (20-gallon, 76-L) vivarium, but it is much safer to house them individually. Forest scorpions seldom breed in the vivarium.

Be aware that forest scorpions often are more nervous and unpredictable than are emperor scorpions and should not be handled.

Flat Rock Scorpions, *Hadogenes* Species Family Hemiscorpiidae

The nearly twenty species of these distinctive scorpions are all found in southern Africa, and most sold in the hobby are called *Hadogenes troglodytes*, which may not always be correct. As far as known, all the species of the genus can be kept the same way and have a relatively low toxicity.

The bodies of these dark brown to blackish flat rock scorpions with paler legs are wide and flattened, in keeping with their habit of hiding in rock crevices. The tail is extremely thin, with long, narrow ring segments that may be almost twice as long in the male as in the female. The male of *Hadogenes troglodytes* often is said to be the longest scorpion at 7.5 inches (19 cm). But, it weighs only a fraction of what a large emperor or forest scorpion weighs—which are just a bit shorter at about 6 inches (15 cm). The pincers are relatively smooth, with blackish fingers that are bent distinctly inward and flattened, and there may be a large triangular knob on the pedipalp patella.

The thin body and narrow pincers of this South African flat rock scorpion (*Hadogenes troglodytes*) allow it to easily slip into rock crevices.

These are dry habitat scorpions that like to squeeze in between stacked rocks (use silicone aquarium sealer to hold the rocks together) and similar tight crevices. They seldom burrow and do well at moderate temperatures and at 60 to 70 percent humidity. Like other scorpions, they like to eat living insects. House specimens singly. Though they can be handled (most are not nervous, and the sting is mild), this is not suggested.

Giant Hairy Scorpions, *Hadrurus* Species Family Caraboctonidae

Giant hairys are the only scorpions from the United States that are regularly found in the hobby. They are big (often 5 inches, 12.5 cm, long), have slender pincers but a thick tail, and have many long bristles on the legs, telson, and front of the carapace (hence the common name: giant hairy). There are six *Hadrurus* species—all burrowers in dry habitats (usually deserts) and found only in Mexico and the southwestern United States. Because of Mexico's restrictions that do not allow exportation of the country's animals, the most commonly sold species in the United States is the Arizona giant hairy (*H. arizonensis*), due to both its popularity and availability.

Generally, the Arizona giant hairy is dark brown over much of the body with a wide, bright yellow triangle in front of the major eyes. (One subspecies is yellow over most

This *Hadrurus arizonensis* is ready to attack even though chewing a cricket.

of its body.) Two other species occur in the United States but are brown to black on the top of the body, and are not nearly as popular.

Humidity will kill off giant hairy scorpions—they must be kept dry (50 to 60 percent humidity) and allowed to burrow in a sandy substrate. If you live in a perpetually humid area, expect them to live short lives in your vivaria, even with the best of care. The substrate should be fairly deep (3 inches, 7.6 cm, or more) and the temperature on the warm side—85°F (29.5°C) or better during the day, dropping five to ten degrees at night. Many specimens remain shy and may not eat. This species also has a reputation for being almost impossible to breed in the vivarium.

Though giant hairy scorpions have slender pedipalp pincers, they have a weak sting; yet they often will sting several times in quick succession before running away, so you should not handle them.

Recently, relatives of the giant hairys have been imported from Peru and Ecuador. These are species of the genus *Hadruroides*, which might be called the "little giant" scorpions. These are smaller scorpions, often less than 3 inches (7.6 cm) long, and different species have different pedipalp pincers. The most commonly imported species at the moment, *Hadruroides charcasus*, is almost uniformly pale yellow-brown in color and has greatly inflated, bulbous pincers. These scorpions can be kept much like giant hairys can, but at a bit higher temperature—about 90°F (32°C), sometimes reaching 95°F (35°C) only for short periods. *Hadruroides charcasus* has been bred successfully in captivity several times, according to reports.

Brown Bark Scorpion, *Centruroides gracilis* Family Buthidae

Because of the brown bark's speed and readiness to sting, you might consider this scorpion more suitable for intermediate keepers than for beginners. But brown barks are widely available and popular in the United States and, if handled cautiously or not at all, make interesting and hardy pets for beginning and advanced keepers alike.

Though hardy and well-suited for the home vivarium, beware of the brown bark scorpion's skill as a master "escape artist."

These are large bark scorpions, sometimes up to 4 inches (10 cm) long, with a slender body and slender pincers. Typical specimens are dark brown over the entire body, with traces of a pale yellowish pattern of spots and dashes on the back, and pale yellow-brown legs. Males are much more slender than females and have a conspicuously longer, narrower tail. There is a strong tooth under the base of the sting (typical of most species of this genus). Brown bark scorpions are found in Mexico and Central America, as well as the Caribbean, and appear to be native to the southern half of Florida, too. Most specimens sold today come from Florida.

These are among the easiest of scorpions to keep in vivaria, as a large colony can be established in a vivarium of only a few gallons (about 11 L) with a thin layer of potting soil on the bottom and stacks of bark for the scorpions to hide in. Similar to other bark scorpions, they cling to the undersurfaces of the bark and don't burrow. Keep them at moderate temperatures and moderate to high humidity.

A word of caution: These scorpions are excellent climbers and "escape artists." They'll take advantage of any escape routes you give them, including piling on top of each other to create an "escape ladder" that reaches up to the edge of the vivarium. Once out, escapees have the troublesome habit of roaming across bedroom ceilings at night, looking for prey, and accidentally falling onto sleepers! They can sting quickly and repeatedly when disturbed. So, take great care to prevent escapes; double-caging is strongly suggested. (See box in chapter 3.)

Though stings from brown bark scorpions from Florida are considered harmless (yet painful), specimens from the Caribbean and parts of Central America have been accused of causing human deaths or at least serious problems. It is uncertain whether the toxicity of this species' venom actually varies among populations from different countries, or whether the reports of fatal brown bark stings are being erroneously attributed to this species.

The striped bark scorpion, *Centruroides vittatus*, is found from the Mississippi River west to New Mexico and Colorado, and north to Nebraska. It often is self-collected. It prefers fairly dry habitats in pinelands and prairies and also is found in deserts from central Texas and westward. Adults often are more than 2 inches (5 cm) long, with males possessing distinctly longer, narrower tails than females have. The body is reddish or yellowish brown with two wide, nearly black stripes down the body, and the area from the major eyes to the front of the carapace is black. Though it has a painful sting, it traditionally has been considered harmless (though recently some dangerous stings have been attributed to this species).

The striped bark does well in a simple vivarium at moderate temperature and humidity, and it breeds quickly and will form large colonies. Remember: beware of escapees—this species moves fast and can disappear into the tiniest cracks and crevices.

Species for Advanced Keepers

Several dangerously venomous scorpions are sold in the hobby but certainly are not suitable for keeping by any but the most advanced keepers. They are just as dangerous as highly venomous snakes, and they can easily escape from their vivaria.

The following species or species groups can kill a healthy adult human if antivenin is not available—and you won't find it at your local pharmacy or emergency room. Generally, sale and transportation of these scorpions across state lines is illegal, and certainly they should never be kept in a home with children. Always keep these species in double cages that are locked and clearly labeled as to their specific venom. (See box in chapter 3.)

Fat-Tailed Scorpions, *Androctonus* Species Family Buthidae

More than a dozen species of this genus are recognized, ranging from the deserts of North Africa across the Middle East as far as India. Though the toxicity of some has been questioned, several are among the most venomous animals known, and at least two species are commonly imported and offered in the pet trade.

These are large (about 4 inches, 10 cm) scorpions with almost square tail segments that usually look like they have been pinched in on each side of the segment top, leaving a narrow ridge down the center. The pedipalp pincers are weak, with long fingers. Overall coloration is yellowish to tan, but the most commonly imported species—the black-tipped fat-tailed scorpion, *Androctonus australis*, from North Africa and the Middle East—is distinguished by a dark brown telson and fifth tail segment. In the other commonly seen species, *Androctonus amoreuxi*, from the same area, the last segment and telson of the tail are yellow like the rest of the body.

Though this species is one of the most deadly, these are hardy scorpions that are easy to keep in a desert vivarium with sandy soil for burrowing, at moderate temperatures, and at about 65 percent relative humidity. They also are among the few scorpions that can be bred consistently in the vivarium, though the young are cannibalistic and must be separated as soon as they leave their mother's back.

Death Stalker Scorpion, *Leiurus quinquestriatus* Family Buthidae

Though a rather pretty, bright yellow scorpion that is hardy and easy to keep, under no circumstances should a death stalker be kept in a home setting. At about 4 inches (10 cm) long, these relatively slender scorpions are quick moving

Appropriately called the death stalker (*Leiurus quinquestriatus*), this scorpion is considered the most venomous of all.

and escape easily from careless keepers. They are even more dangerous than fat-tailed scorpions, and definitely can kill even a heavy adult human. This scorpion has few distinguishing features, though often the last (fifth) ring of the tail and the sting are dark brown, contrasting to the yellow of the rest of the tail. The pedipalp pincers are weak, with very long fingers, especially in males, which typically have longer, thinner tails than females have. Often there are long bristles on the walking legs and tail. This is a common species from northeastern Africa across the Middle East, where it lives in deserts and dry habitats. Death stalkers can be kept in a dry vivarium much like fat-tailed scorpions.

Thick-Tailed Scorpions, *Parabuthus* Species Family Buthidae

This is a vast (more than two dozen species), confusing genus of generally large (4 to 5 inches, 10 to 12.5 cm, long) scorpions found in dry areas of southern and eastern Africa to the Arabian Peninsula. In shape, they are much like fat-tailed scorpions, having squarish tail segments and slender pincers, but coloration is much more variable, ranging from golden yellow to solid dark brown. These hardy scorpions live in burrows under rocks and logs in deserts and dry savannas. Unfortunately, at least some species (many are rare and virtually unknown) are dangerously toxic, even deadly. These scorpions have an interesting wrinkle in their

The granulated thick-tailed scorpion (*Parabuthus granulatus*) is considered the most venomous species in its native habitat in South Africa and Namibia. This species is not known to spray its venom as some of its relatives have the ability to do.

venom delivery system: they can spray their venom toward a predator from distances as much as 3 feet (91 cm). When sprayed, the venom can cause intense pain and temporary blindness if it gets into the eyes. Scorpions view keepers as predators, so safety goggles are a must when these scorpions are kept as specimens. The most commonly offered species is *Parabuthus transvaalicus*, a large blackish brown scorpion noted for its venom-spraying behavior.

Thick-tailed scorpions can be kept much like fat-tailed scorpions. They have been bred in the vivarium on occasion, with mating behavior stimulated by misting to simulate rains after a dry period.

CHAPTER 7

WHIP AND WIND SCORPIONS

People interested in keeping scorpions as pets will probably also be interested in the very strange arachnids known as whip scorpions and wind scorpions. Though not that closely related to scorpions, they have a similar general appearance and never fail to attract attention when displayed.

Whip Scorpions

Members of two different groups are widely called whip scorpions because of their greatly elongated first pair of walking legs used to help sense prey. The order Uropygi contains the species known as tailed whip scorpions, and the members of the order Amblypygi are called the tailless whip scorpions or whip spiders. The two types of animals are not that similar except for the long first legs, but both have large pedipalps that are used to capture prey (usually, small insects and other invertebrates; and, rarely, frogs and lizards) and rip it apart with the help of the chelicerae, much as in scorpions.

Tailed Whip Scorpions

Uropygids look much like scorpions at first glance, having a large carapace and a rather swollen abdomen. At the end of the abdomen is a short "tail" of three segments and then a threadlike flagellum that may be longer than the abdomen. Most tailed whip scorpions are 1 to 3 inches (2.5 to 7.6 cm) long (not counting the flagellum, which often breaks off) and are dark brown in color. The flagellum, surprisingly, is

With its odd-looking, whiplike abdominal flagellum, the common tailed whip scorpion (*Mastigoproctus giganteus*) makes an interesting-looking pet for beginning and advanced keepers alike. Whip scorpions are not venomous, but the tailed varieties produce a sour-smelling spray that serves as their main defense against predators.

light-sensitive, though it cannot detect images. There are four pairs of walking legs having the same segments as in scorpions, but the segments of the first pair of legs are greatly elongated and serve as antennae to help sense prey and predators. The pedipalps are large and thick and don't end in pincers as they do in scorpions. Tailed whip scorpions pounce on their prey, grabbing it with the pedipalps before transferring the food to the chelicerae to be ripped apart for digestion. Their eyes are strange, with a pair of eyes near the very front of the carapace and a group of four or five eyes much farther back on each side of the carapace placed approximately over the bases of the second legs.

Tailed whip scorpions often are called vinegaroons because they have a pair of glands (repugnatorial glands) at the end of the abdomen that can shoot out a spray of organic acids, including acetic acid (vinegar). Not only does this mixture stink, if it hits a predator (or keeper) in the face, it can cause vision problems, headaches, and nausea. Vinegaroons lack venom glands, so their spray serves as their main protection. Their bite (whether from the pedipalps or the chelicerae) is harmless, though a large uropygid can draw blood.

Common vinegaroons have interesting mating habits. Males court females in much the same way as scorpions do, and they also deposit a spermatophore containing a sperm packet onto a hard spot on the substrate. The male guides the female near the spermatophore, and then uses the modified tip of his pedipalp to pick up the spermatophore and push it into the genital opening on the belly of the female. The female develops a small sac over the genital opening and deposits her fertilized eggs into the sac, which she carries around on her belly for about a month as the eggs develop. It may take as much as four years for a vinegaroon to reach sexual maturity, after which it may live another three or four years. In some uropygids, the mother dies shortly after the eggs hatch and the young leave her.

About ninety species of vinegaroons are found in the tropics of the Americas and Asia, with only one likely to turn up in the hobby. This is *Mastigoproctus giganteus*, the common vinegaroon, a dark brown uropygid found in Mexico, Central America, and the Caribbean, as well as in Florida and the Southwest in the United States. A highly adaptable but generally uncommon and secretive animal, common vinegaroons can be found under debris on the ground in both humid areas and dry, even desertlike habitats. They feed on a variety of small invertebrates and readily take crickets and superworms in the vivarium. If given a moderately warm vivarium with a substrate of potting soil and cover in the form of a piece of cork bark or a few rocks, they may live quite well. Keep the humidity level near 75 percent.

Tailless Whip Scorpions

These arachnids are actually closely related to spiders (thus the common name, whip spider, used by Europeans), but they lack venom glands in the chelicerae and have a broad, flattened shape and a strongly segmented abdomen. The pedipalps are very strong for a small animal (most amblypygids are half an inch to 2 inches, 1.25 cm to 5 cm, long) and are armed with long spines that impale as well as crush the prey (and can produce a painful, bloody pinch if you are careless). Amblypygids lack a distinct tail or flagellum as found in uropygid whip scorpions. Their crowning glory is the elongated first pair of walking legs, which in a species with just 2 inches (5 cm) of body length may be 10 inches

(25 cm) from tip to tip. The long legs serve as antennae to sense prey when the arachnids hunt at night under leaves, bark, and stones. They have a characteristic crablike, sideways walk that makes them instantly recognizable.

Amblypygids are common in the tropics of the world, with at least seventy-five described species. A few species are found in the United States' Southwest under near-desert conditions, and one or two species also occur in humid southern Florida. The largest species are from tropical America and southern Asia. Often imported are large species (some almost 2 inches, 5 cm, in body length) that may belong to the genera *Paraphrynus* (pedipalps ending in a simple tip) and *Damon* (pedipalps ending in a pair of fingers). Much more commonly seen is *Tarantula marginemaculata*, a half-inch (1.25-cm) species with rows of yellow spots on the abdomen and long spines on the pedipalps. This is the common species of southern Florida and makes a hardy pet.

Mating takes place much as it does in scorpions and uropygids, the male depositing a spermatophore on the substrate and leading the female over it so she can pick it up in her genital pore. A flattened sac develops behind her legs to hold the eggs and developing young, which may be carried about by the mother for three months or more. Amblypygids grow slowly and molt about annually, even after maturing.

The long legs of the tailless whip scorpion help the animal find prey hiding under bark, rocks, and leaf litter. Instead of the odorous spray defense mechanism of tailed whip scorpions, tailless varieties are fast moving and prone to pinch.

Tailless whip scorpions can be kept in much the same fashion as tailed whip scorpions and are active at night. Feed them on crickets and other invertebrates of appropriate size. They can be long-lived little animals.

Wind Scorpions or Solpugids

Truly the weirdest of the weird, the wind scorpions—also called solpugids—are arachnids that are not closely related to any of the other groups we've talked about so far. Known by many names—including camel spiders, sun spiders (because they often are active during the heat of the day), and solifugids—a few large species as much as 3 inches (7.6 cm) in body length appear for sale, usually imported from the deserts of Africa and the Middle East. The names *Galeodes*, *Rhagodes*, and *Zerassia* often are applied, but possibly are incorrect. Of more interest to scientists than to hobbyists, solpugids have a very poor record in the vivarium and seldom survive more than six months (usually just a few weeks), perhaps in part because adults are naturally short-lived. There are close to one thousand species found in tropical and semitropical areas around the world (except Australia), with more than two hundred species (all small) recorded from the deserts of the southwestern United States and one small species in southern Florida.

Solpugids are more strange-looking than the other arachnids, at least in my opinion. They have a large, oval,

Solpugids are not poisonous, and many specimens such as this green sun spider are quite attractive. Their short life spans and high maintenance requirements, however, leave them poorly suited for the home vivarium.

This sun spider, spotted in northern Tanzania, consumes a grasshopper.

strongly segmented abdomen without a tail, and the four pairs of walking legs originate under a rather small trunk. The head and trunk (collectively called the prosoma) are divided into two parts and bear a pair of large black eyes at the front. What you will notice first is the pair of gigantic chelicerae, which are about as long as the trunk and may in some species be a third of the total body length. The chelicerae consist of a greatly inflated fixed or upper finger and a smaller movable finger underneath, both bearing a variety of teeth and ridges used for grabbing and ripping prey. There also are many specialized bristles on the chelicerae, sometimes including a featherlike flagellum used by the male during mating.

The pedipalps are long and leglike, ending bluntly in an adhesive organ that is used to help catch prey and also to climb. The first pair of walking legs are slender and often are held up and out in front of the body, serving as antennae to sense prey. Male solpugids have longer legs than females have, and they are swift runners (running like the wind, thus the common name) and climbers, whereas females generally are more subdued. Most solpugids are small, less than an inch (2.5 cm), but several tropical species are giants at nearly 3 inches (7.6 cm). In many wind scorpions, there are long bristles over the body and legs, and some species look almost as hairy as tarantulas. Colors vary from black or gray to bright reddish brown, seldom with strongly developed color patterns.

Solpugids often live in deserts and dry savannas, where they may be active both day and night, feeding on any small animals they find, including lizards and frogs. Some species are found in tropical forests, and a few even live in termite mounds. Termites, in fact, are a favorite food of almost all solpugids. Solpugids are among the few animals that can survive direct desert sunlight.

Breeding is rather odd. Because females are likely to kill and eat males upon sighting, it also is a bit tricky for the male. He sneaks up on a female and caresses her abdomen and prosoma with his legs, which seems to paralyze her. Then he moves her to a protected spot and continues to massage the tip of her abdomen while he releases sperm and transfers it to the female's genital opening with his chelicerae. As soon as sperm transfer is complete, the male moves away and the female recovers. Shortly thereafter, she digs a hole under a log or rock and lays a few dozen to more than two hundred eggs. She may stay with the eggs as they develop, or leave them to their fate. Hatching may take a few hours to several weeks, but in either case the eggs produce small, white young that cannot feed until after their first molt (as in scorpions).

Hobbyists have had few long-term successes keeping solpugids, with either the common desert species of the United States or the large imported species. The usual suggestion is to keep them in fairly roomy vivaria according to their size, and to provide them with a substrate of sandy soil and a few rocks or pieces of bark under which to hide or burrow. Temperatures can be quite warm (as much as 90°F, 32°C) during the day, dropping about ten degrees at night. Handle the animals as little as possible (they are easily damaged) and feed them on crickets and other small invertebrates. Solpugids lack venom glands of any type, so their bite is harmless; however, large species can draw blood.

CHAPTER 8

MILLIPEDES AND CENTIPEDES

Hobbyists often confuse millipedes with centipedes (sometimes spelled millipeds and centipeds) since both are greatly elongated animals with a pair of antennae and a great many legs. The two groups are together called the class Myriapoda and are somewhat ancestral to the insects, sharing many features of their structure. Unlike the arachnids we've talked about so far, both have true jaws and antennae, and they lack pedipalps. The food is actually eaten, the pieces being chopped up by the jaws and swallowed to be digested in the body. Both are very large groups of animals with a worldwide distribution, though only the largest species are of interest to hobbyists.

Keeping Millipedes

Millipedes are unique among arthropods (animals with jointed legs) in having two pairs of legs on each apparent segment of the body. This is the origin of their subclass name, Diplopoda. All millipedes are elongated, with at least eleven segments in addition to the head and more often between twenty-one and sixty total body segments. No millipede actually has a thousand legs (as implied by the name); the typical range is between nineteen and one hundred pairs—the largest recorded number is 375 pairs. The legs move in synchronized waves that are amazing to watch. Except for a few species, they have strong jaws and are vegetarians, feeding on dead leaves, fungi, and fine rootlets; some actually attack garden vegetables and are considered economic pests. Millipedes live in the soil or under debris on the forest or even desert floor, though some are excellent climbers. They are active mostly at night. The large species found in the hobby have compound eyes (a cluster of tightly placed simple eyes as in insects) behind the antennal bases, but many other millipedes lack eyes.

To mate, millipedes move along in parallel for a while and then turn so their bellies are next to each other, holding on to each other with their many legs. The male moves forward so his seventh segment, which contains legs modified for sperm transfer, is opposite the female's second segment and then deposits packets of sperm into the female's sperm receptacles at the bases of the second legs. Pairs may remain joined for hours. Later, the female digs a small nest hole in the soil and deposits many round eggs that hatch into tiny white larvae that have many fewer segments than seen in adults. Each egg is embedded in a fecal pellet that provides it with a stable environment and probably a first meal for the hatchling. With each molt the young add segments and legs until the final molt to adulthood. Commonly, millipedes die after mating and laying eggs; adults of small species probably have life spans of just one to three years.

The millipedes of pet interest tend to be the largest species (those more than 4 inches, 10 cm), and all these have bodies made from nearly circular rings, giving them an elongated, tubelike appearance. Most of the big, round millipedes are reddish or shades of brown, but some have attractive patterns of red spots or lines on a black background.

Millipede Vivaria

Almost all common millipedes are soil animals that do moderately well in a small vivarium (a gallon, 3.8 L, or so) with a deep layer of crumbled leaf mulch and soil along the bottom. Start by putting an inch (2.5 cm) of fertilizer-free potting soil or garden soil in the bottom of the container and then cover this with another inch (2.5 cm) of coarsely chopped leaf litter from your yard. Millipedes feed on the composted leaves as well as on the fungi and algae growing on the leaves, even when fed lettuce and other common foods. Small millipedes may disappear into the substrate, but larger species will usually spend the day under pieces of bark at the surface. Millipedes dislike light (though many lack eyes, they still can sense light), high temperatures, and very dry conditions. Make sure the substrate stays moist (but never wet) by adding small amounts of water to the corners of the vivarium, or by misting daily. This simple caging will work well for most millipedes.

Desert Millipedes

The genus *Orthoporus* contains at least fifty species of fairly large millipedes found in a variety of habitats from Texas in the United States to Brazil, but only one species is commonly sold as a pet. This is the western desert millipede, *Orthoporus ornatus*, a yellow to reddish brown species found from Nevada to Texas in deserts and dry prairie habitats. Adults commonly are 6 inches (15 cm) long and about half an inch (1.2 cm) in diameter, with reddish brown rings alternating with yellow rings; some specimens are entirely yellow, others entirely reddish brown. Much of their life is spent underground in relatively humid rodent burrows or root holes, but they come to the surface when desert rains provide a cooler, more humid climate; at this time they can be seen crawling across roads, where they are collected in large numbers for sale. Colorful, large, inexpensive, and hardy—they are great millipedes for beginners.

Narceus americanus is a large, brightly colored round millipede that is common over much of eastern North America and makes a good pet. It can be kept in a humid vivarium with leaf litter.

Three or four desert millipedes can be kept in a 1-gallon (3.8-L) vivarium with a simple slotted top and about 2 inches (5 cm) of sandy potting soil on the bottom. Give them a piece of bark to hide under and a shallow water dish. (Like some other millipedes, desert millipedes can drink by absorbing water through the anus as well as through the mouth.) They like to climb and will use a low climbing branch. Keep the temperature moderate, about 79°F to 86°F (26°C to 30°C), dropping a bit at night, and the relative humidity at least 70 percent. Mist the vivarium once or twice a week. Feed them on green lettuce, shredded carrots, zucchini, apples, and almost any other firm fruit or leafy green. Mating sometimes occurs in the vivarium, though the young seldom survive.

Most other large millipedes from almost any part of the world can be kept in much the same fashion and will do at least fairly well for a year or more.

Giant African Millipedes

These are the millipedes that draw attention in any shop: 10 inches (25 cm) long and an inch (2.5 cm) in diameter, with most specimens a deep blackish brown with reddish legs. Traditionally, they are called the giant African millipede (*Archispirostreptus gigas*), the name of a species from central and eastern Africa. However, since many specimens are imported from western and perhaps southern Africa, there probably are several species in the hobby today. All have similar pet qualities, and dealers often sell them mixed together.

Giant millipedes are great pets, especially if you enjoy the strange feel of dozens of small legs moving in waves across your hands and arms (which I do). They love to climb. Their feet can find the smallest irregularities in the sides of a plastic vivarium and, if there is no lid, they soon will fall to the floor. Fortunately, they seem to be resilient and usually survive even long falls unscathed, wandering off looking for food. This means that you want to give them a large (at least 5 gal, 19 L), tall vivarium with a sturdy screen cover that can be latched into place—a determined giant millipede can be quite strong.

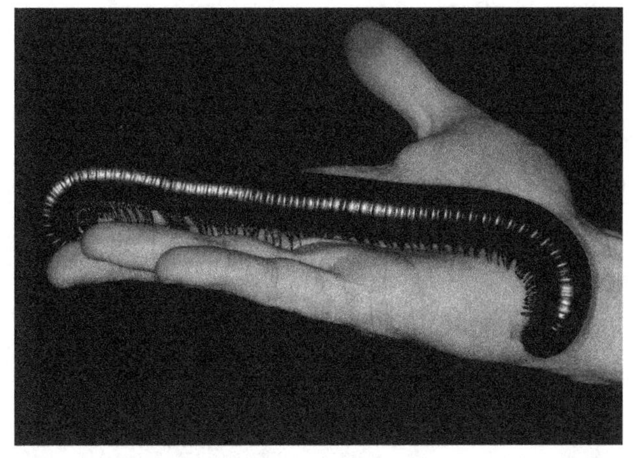

At nearly a foot (30 cm) long, the giant African millipede certainly lives up to its name!

A couple of inches (5 cm) of potting soil does well as the substrate, with a shallow water dish and a climbing branch or two to complete the decor. They feed on large amounts of lettuce and other greens and some fruits. Mist the vivarium every day or two to maintain a humid atmosphere, and keep them at moderate temperatures. Special lighting is not needed, though they are not shy about normal room lights. Many giant millipedes live three or four years, sometimes producing tiny young that grow well in the vivarium with their mother. These are quite social millipedes, so several can be kept in a large vivarium without problems. Males seldom are imported, but many imported females appear to be gravid.

One minor problem with giant African millipedes is that they often carry large numbers of small, flat, light-brown mites. These mites feed on leftover bits of food on the mouthparts and legs of the millipedes and roam freely over their bodies looking for food. A single millipede may have a dozen or more mites. Fortunately, these are commensal mites (living at the table of another animal) and not parasites, so they are harmless to both you and the millipede. I prefer to just leave them alone, but they are kind of creepy if they leave the millipede and wander across your hands while you are handling your pet. Chemical treatments formulated to kill the mite will also kill the millipede; so, if you want to remove the mites, you'll have to pick them off individually with fine tweezers or, better yet, a toothpick dipped in petroleum jelly.

Keeping Centipedes

Keepers of scorpions often share their attention with centipedes. This probably is because all the large centipedes are venomous animals, some capable of inflicting a dangerous bite rumored to have killed humans (though this is hard to verify). Centipedes have many of the same attributes as scorpions: they can be housed in similar ways, feed on similar foods, and are great escape artists. Some also are colorful animals.

Centipedes also are myriapods, belonging to the subclass Chilopoda. They have at least nineteen segments in addition to the head and, in some very slender burrowers, may have more than 150 total body segments. Usually, each body segment bears a pair of legs, with the last pair enlarged and often flattened and toothed and carried in line with the axis of the body. This last pair can be used to grasp prey, and perhaps can be considered a faux head used to confuse predators. The head has a pair of antennae and jaws and, usually, a few simple eyes. The fangs of a centipede actually are the modified first pair of legs, which are thickened at their bases and each end in a hollow claw containing the

Centipede fangs—on this specimen, seen just behind the antenna—function more like pincers than like true fangs to deliver venom. Centipedes use these "pincers" to seize prey and inject it with venom that comes from glands inside the fangs.

opening of the venom glands. Technically, the "bite" of a centipede is a venomous pinch. The hind legs, contrary to many myths, are just legs, not poison fangs. Most centipedes are greatly flattened; half the shields on top of the segments are squarish and often are separated by narrower shields in an alternating fashion.

Centipedes are difficult to sex because their sexual organs are housed in a small segment at the very back of the body between the hind legs. Because centipedes often are cannibalistic, the male cautiously approaches a female and thumps on the ground with his antennae to get her attention. If she doesn't escape or try to eat him, he continues to tap his antennae along her body for an hour or more and then gets to business. He spins a small web near the female and deposits a sperm packet (spermatophore) on it, then moves away. The female walks over the web and picks up the sperm packet with her genital segment at the end of the body. Shortly thereafter, she digs a nest hole in which she curls up. The eggs are large and pearl-like, held within the coil of the female so they don't touch the soil. The mother guards the eggs against small predators and fungus, keeping them moist and clean. The eggs hatch in one to two months, and the young will molt several times before leaving the nest chamber. Centipedes, even the large species, mature quickly, usually in six months or so, and can live four to six years.

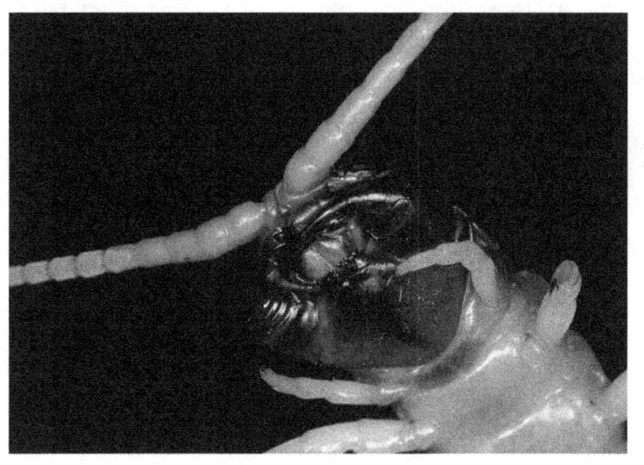

This close-up view of a centipede's underside shows the fangs and mouthparts.

Centipede Vivaria

Centipedes are soil animals, with most of the smaller species living under debris or in tunnels and holes in the soil. Larger centipedes, those of interest to hobbyists, often are climbers, residing under bark and in soil accumulations in root masses of smaller plants living on tropical trees. They require very little to live well in a suitable vivarium, regardless of their origin.

Hobbyists are generally interested only in large centipedes, those reaching at least 4 inches (10 cm) long with the largest, regularly available species reaching 10 inches (25 cm). Centipedes are active animals, so they need a roomy cage; because they are escape artists and venomous, they require a tight-fitting lid that is at least partially screened and that can be latched into place without leaving narrow cracks at the edges. You can use a plastic critter box if you place aluminum screen under the slotted lid. Give a large centipede at least 1 gallon (3.8 L) of space. An inch or two (2.5 cm to 5 cm) of potting soil with some live peat moss mixed in serves as a good base, especially if covered with pieces of bark to help hold humidity.

The temperature should be moderate to warm, about 86°F (30°C), and the humidity about 70 percent. Mist the vivarium every other day. Centipedes usually don't like bright lights. Though they don't need decorations, provide a branch for climbing and a shallow water bowl. Clutter in the vivarium makes it easy for a centipede to escape. Most centipedes will take crickets, superworms, other invertebrates, and even the occasional frog, lizard, or newborn mouse—and other centipedes.

Some Giant Centipedes

Identification of centipedes is difficult to impossible, but all the species of interest to hobbyists belong to the family Scolopendridae, and usually to the genera *Scolopendra* or *Hemiscolopendra*. Centipedes in a great variety of colors are sold from sometimes dubious localities, and little is known of the variability within giant centipede species.

Ranking second in length among the giant centipedes, the Vietnamese centipede (*Scolopendra subspinipes*) is a stunning sight. Its bright red legs (one variety has yellow legs) stand out in sharp contrast to its black body. *S. subspinipes* is known to attack quickly and is a voracious eater.

From North America we are likely to see:

- *Scolopendra alternans*, an 8-inch (20-cm) species that is uniformly yellow on the top of the body;
- *Scolopendra heros*, a 6-inch (15-cm) or longer species with a reddish head, blue-black posterior segments and last legs, and yellow to blue-black or green trunk segments, set off by yellow legs;
- and *Scolopendra polymorpha*, a rather small (4 inches, 10 cm) species that is dark reddish brown with black stripes across the front edges of the segments.

Tropical species include:

- *Scolopendra morsitans*, an 8-inch (20-cm) dangerous species brightly striped in yellow and green;
- *Scolopendra gigantea*, a 10-inch (25-cm) generally yellowish species;
- and *Scolopendra subspinipes*, an 8-inch (20-cm) giant with a wicked-looking pattern of a black body and head and red legs.

CHAPTER 9

STRANGE INSECT PETS

There are more insect species than all other species of living animals combined (estimates range from one million to forty million species), yet few are kept as pets by American hobbyists. Perhaps this is because of stringent agricultural laws prohibiting random shipments of insects across state lines. Even to ship food crickets and mealworms, a commercial dealer has to comply with a sheaf of local, state, and federal regulations.

Only three groups of insects are likely to be sold as pets today, all close relatives of the grasshoppers and crickets. Mantises, roaches, and stick insects share a type of development where the egg hatches into a tiny, wingless version of the adult—the nymph—that molts many times before reaching adulthood. With each molt, resemblance to the adult increases until, in the final molt, they attain functional wings (if the species has them) and become sexually mature.

Mantises

Mantises are famous for females eating their mates (which does happen, but not always) and for being highly predaceous, with the front legs bearing spines and cutting edges that can help catch and kill prey heavier than themselves. Though mantises usually prey on flies, wasps, grasshoppers, and such, they have been seen to catch frogs and salamanders and the occasional lizard as well. In a mantis, the body is elongated, long wings often are present, the front legs are modified for catching prey, and the head has large compound eyes and can swivel freely to "watch its back." Most mantises (insect order Mantodea with nearly two thousand species) are large for insects, often 2 to 4 inches (5 to 10 cm) long, and vary in color from plain browns and greens to highly complex patterns and brightly colored hind wings.

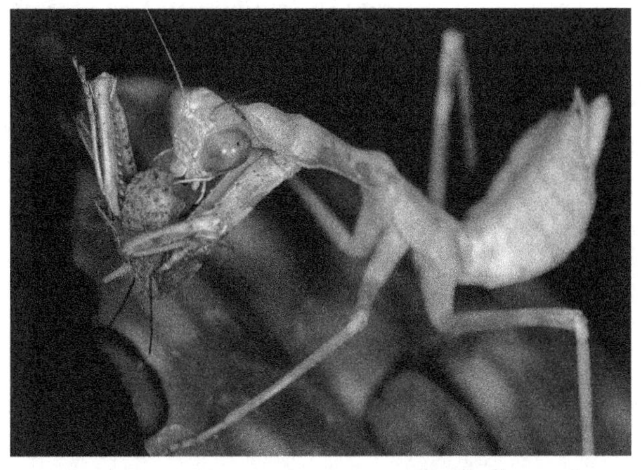

A mantis makes a meal of a passing cricket.

Females have a short ovipositor and generally are larger than males. They lay many eggs encased in a tough case surrounded by a froth that dries into a woodlike consistency. Egg cases usually are deposited onto a hard substrate and hatch in a few days to weeks. Young mantises are highly cannibalistic and if not separated immediately will begin consuming their siblings.

Mantises are found almost everywhere, from deserts to tropical forests, and occur on every continent. They often are inconspicuous animals with colors that blend into the background, but they are numbered among the major groups of micropredators. Few adult mantises live more than one season, which means that a hobbyist has to maintain a breeding

The ghost leaf or alien mantis (*Phyllocrania paradoxa*)—takes the cake when it comes to strange looks.

Although some mantises are designed to blend in with their woody and arboreal habitats, the bullseye mantis certainly stands out.

program to keep a line going for very long. Some mantises are sold in horticultural catalogs because of their helpful predation on insect pests of flowerbeds and gardens. The most common mantises seen in the United States are actually introduced species from Europe and Asia.

If you buy or collect a mantis, try to get an early stage before the wings are developed, which will help assure that you have your pet for at least a few months. Keep them in plastic critter boxes with slotted lids (or even large glass jars; drill air holes in the top). A light layer of potting soil substrate on the bottom, a thin top layer of leaf litter, and some climbing branches complete the decorations. Mist the vivarium daily (the cage, but not the insect) and also provide

Since mantises have short life spans, hobbyists looking for a pet should try to select a juvenile.

a bottle cap as a water bowl. Small mantises need small cages so they can easily find their food. For their own safety, house all mantises in separate vivaria to prevent cannibalism.

As food, try pinhead crickets for baby mantises and adult crickets for adult mantises. They also take a variety of flies, cultured or wild-caught (beware of pesticides), and other soft-bodied insects that can be found in fields and yards. Don't forget caterpillars and small moths.

Breeding involves putting two specimens together and risking the male; in some species, full fertilization only occurs after the male's head is ripped off, the dead male continuing to pour sperm into the female.

Roaches

Forming the order Blattaria with more than thirty-five hundred species placed in as many as twenty-eight families, roaches are familiar to everyone as nasty pests. However, there are many attractive and nonpesty tropical roaches that can be kept in the vivarium and will freely breed. Some are kept just as pets—but admittedly, most are kept to supply other vivarium animals with food.

Roaches bring out the hunter in many other pets, and lizards such as chameleons go wild for them, especially green-colored species. All the roaches look much the same: flattened, rather oval insects with a large first segment that commonly covers all or part of the head; strong, spiny legs;

Gromphadorhina sp. roam along a wood branch.

Note the sensory bristles on this colorful roach's legs.

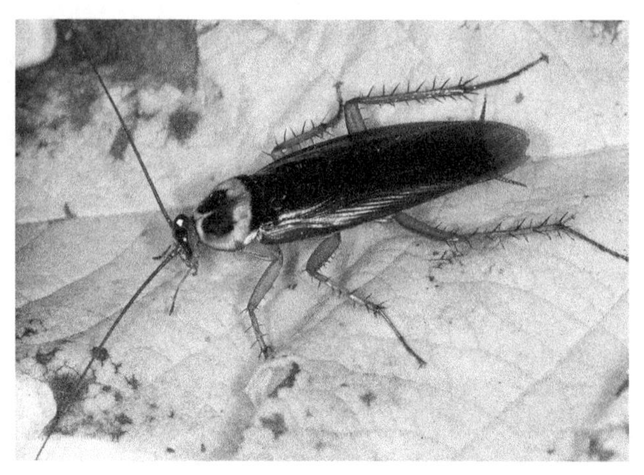

and the first pair of wings leathery (but many roaches lack wings entirely). Females produce an egg case called an ootheca that contains dozens or hundreds of eggs. Nymphal roaches often are almost round, lack wings, and often lack the interesting color patterns of the adults. Many larger roaches (including the most common pets) give live birth by holding the ootheca inside the body until the eggs are ready to hatch.

Roaches do well in glass or plastic cages with tight-fitting lids that are part screen and part glass (to increase the humidity). Commonly, they are given a substrate of shredded aspen or pieces of bark and a variety of hiding places on the floor. Feed them on a variety of grains (oatmeal, stale bread) and bits of apple or other hard fruits as well as some dry dog food. Beware of fungus developing, as the humidity will be quite high and the air movement low. Keep your roaches away from bright lights and at warm room temperatures (80°F, 26.5°C, or more).

If you worry about escapes (some smaller roaches can climb glass, but common pets are not as agile), smear a band of petroleum jelly around the inner edge of the vivarium below the lip. This will keep any roach from crawling out, but it must be replaced monthly. Many roaches fly freely, thus the need for the tight lid.

The most commonly sold "pet" roaches probably are the Madagascan hissing cockroach (*Gromphadorhina portentosa*) and the giant tropical or deathhead cockroach (*Blaberus giganteus* and relatives). Both are large, heavy roaches at least 2 inches (5 cm) long. The hissing cockroach lacks wings, whereas the giant has long wings bearing a dark-brown pattern. Both are kept in much the same way as other roaches, but with warmer temperatures in which to be able to breed. Give them a range from about 82°F to 88°F (28°C to 31°C) but below 90°F (32°C) to assure free breeding. Male hissing cockroaches are distinguished from females by their wide black protuberances ("horns") behind the head; nymphs of the *Blaberus* species are nearly round and often brightly marked with red spots. Both these roaches are too heavy to climb glass easily and the *Blaberus* seldom fly; but the nymphs can be amazing escape artists and turn up in strange places, so be careful.

Stick Insects

Stick insects or walking sticks, order Phasmoptera, are common pets in Europe. They are hard to find in the United States, however, because most species are considered potential agricultural pests and thus cannot be legally transported around the country. Perhaps some day the laws will change.

Stick insects are close relatives of the grasshoppers, but their hind legs are not adapted for jumping, the body is long and narrow, and usually the wings are short or absent. They feed on a variety of plants, with each species often restricted

This giant walking stick (*Diapheromera denticrus)* from Texas measures 10 inches (25 cm) long!

to a narrow group of related plants, and lay eggs that are interestingly sculptured or armored and mixed in with their fecal droppings on the floor of the cage. Eggs often need a cooling period before hatching, and the nymphs may be very particular about the leaves they will eat.

Though there are more than twenty-five hundred species of walking sticks known, mostly from the tropics, only a few dozen are ever kept as pets, usually the largest and showiest forms. Many of these look like thorny twigs, bearing a variety of spines, but some are quite colorful. In most species, the female is much larger than the male; a male may attach to a female's body to continue mating over several weeks.

Because stick insects live short adult lives, it is necessary to breed them to keep a line going. Their care is quite specialized and beyond the aims of this book, but several good books (mostly British) are available on their care. (See Resources.) You can keep a small group of stick insects (they are quite social) in a tall, well-ventilated cage, preferably a mesh cage such as is used for chameleons. A carefully rolled piece of window screen placed in a gallon jar for a base and closed off at the top works well, too. Place a food plant in the cage and mist the enclosure regularly; the insects just eat their way through life.

Hatching or even recognizing the eggs in the bottom of the cage can be complicated. Few stick insects live more than two or three months after becoming adult.

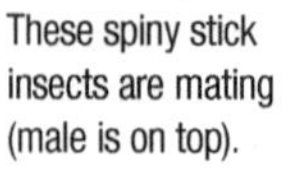

These spiny stick insects are mating (male is on top).

RESOURCES

Societies

American Tarantula Society
PO Box 756
Carlsbad, N. Mex. 88221-0756
On the Web at: http://www.atshq.org
Publishes *The Forum Magazine*

British Tarantula Society
c/o Angela Hale
3 Shepham Lane, Polegate
East Sussex BN26 6LZ England
On the Web at: http://www.thebts.co.uk
Publishes *Journal of the British Tarantula Society*

American Arachnological Society
(a scientific society devoted to all arachnids)
c/o Dr. J. W. Schultz
Department of Entomology
University of Maryland
College Park, Md. 20742
On the Web at: http://www.americanarachnology.org
Publishes *Journal of Arachnology*

Web Sites

http://www.petbug.com
Care sheets and links for many invertebrates.

http://www.science.marshall.edu/fet/euscorpius/INDEX.htm
Online technical journal, *Euscorpius*, devoted to scorpions.

http://www.ub.ntnu.no/scorpion-files/index.php
Constantly updated source of technical information on scorpion families and genera.

Books and Articles

Brock, P. D. *A Complete Guide to Breeding Stick and Leaf Insects*. Neptune City, N.J.: TFH, 2000. Excellent coverage of the intricacies of stick insect care and breeding.

Gaban, R. D. *Gaban's Scorpion Tales*. Carlsbad, N. Mex.: American Tarantula Society, 1998. Compilation of scorpion articles from *The Forum Magazine*. Practical and useful.

Jackman, J. A. *A Field Guide to Spiders and Scorpions of Texas*. Houston, Texas: Gulf Publishing, 1997. Good natural history information.

Keegan, H. L. *Scorpions of Medical Importance*. Jackson, Miss.; University Press of Mississippi, 1980. A bit outdated, but beautiful drawings of many dangerous scorpions.

Parker, S. P., ed. *Synopsis and Classification of Living Organisms*. Vol. 2. New York: McGraw-Hill, 1982. Excellent technical information on almost all groups of arthropods.

Polis, G. A., ed. *The Biology of Scorpions*. Stanford, Calif.: Stanford University Press, 1990. The standard technical reference on all aspects of scorpiology; systematics section (by W. D. Sissom) becoming outdated. Out of print, but look for the title at used bookstores and online.

Preston-Mafham, R. *The Book of Spiders and Scorpions*. London: Quarto, 1991.

Walls, J. G. "Madagascan hissing cockroaches," *Reptile Hobbyist* 2, no. 10: (1997): 68–71.

———. *The Guide to Owning Millipedes and Centipedes*. Neptune City, N.J.: TFH, 1999. The only book on care of common millipedes.

INDEX

ABOUT THE AUTHOR

A native of central Louisiana, **Jerry Walls** worked as an editor in New Jersey for more than thirty years, authoring more than four hundred publications on natural history subjects, especially reptiles and amphibians. His thirty-eight books (including twenty on herps) range from introductory works on lizards and turtles as pets to massive reviews of seashells, boas and pythons, and poisonous frogs. He also edited *Reptile Hobbyist* magazine and currently writes a monthly column for *Reptiles* magazine. He is an active birder, with more than six hundred U.S. species on his life list, and has authored several books and articles on pet and wild birds. Collecting crawfishes, snails, and herps for taxonomic study currently is his favorite preoccupation.

CPSIA information can be obtained
at www.ICGtesting.com
Printed in the USA
LVHW080150070919
630291LV00008B/169/P